The Owl

Also by JOHN HAWKES

Charivari (in *Lunar Landscapes*)

The Cannibal

The Beetle Leg

The Goose on the Grave & The Owl
    (also in *Lunar Landscapes*)

The Lime Twig

Second Skin

The Innocent Party (plays)

Lunar Landscapes (stories & short novels)

The Blood Oranges

Death, Sleep & the Traveler

Travesty

Of related interest

A *John Hawkes Symposium: Design and Debris*, edited by Anthony C. Santore and Michael N. Pocalyko, New Directions

John Hawkes: An Annotated Bibliography; With Four Introductions by John Hawkes, by Carol A. Hryciw, Scarecrow Press

JOHN HAWKES
The Owl

*With a Critical Interpretation
by Robert Scholes*

A New Directions Book

Copyright 1954 by John Hawkes; copyright © 1969 by John Hawkes
Copyright © 1977 by New Directions Publishing Corporation

All rights reserved. Except for brief passages quoted in a newspaper, magazine, radio, or television review, no part of this book may be reproduced in any form or by any means, electronic or mechanical, including photocopying and recording, or by any information storage and retrieval system, without permission in writing from the Publisher.

Robert Scholes's "Introduction" was first presented during the John Hawkes Symposium, Muhlenberg College, April 9 and 10, 1976, and was first published in *The Hollins Critic*, Summer 1977.

Manufactured in the United States of America
First published as New Directions Paperbook 443 in 1977
Published in Canada by McClelland & Stewart, Ltd.

Library of Congress Cataloging in Publication Data

Hawkes, John, 1925-
   The owl.
   (A New Directions Book)
   I. Title.
PZ3.H31320W   [PS3558.A82]   813'.5'4   77-8227
ISBN 0-8112-0665-3 pbk.

New Directions books are published for James Laughlin
by New Directions Publishing Corporation
333 Sixth Avenue, New York 10014

Introduction

For over twenty-five years John Hawkes has been a unique voice in American letters. Belonging to no school, following no fashion, he has paid the price exacted of such loners by the literary establishment. He has been reviewed capriciously, embarrassed by unconsidered praise and attacked with ill-tempered venom. His admirers have been mostly his fellow writers, some English teachers, and their students. Recently he has been more honored abroad than at home. For better or worse, he has been taken up by the French, who can see in his writing connections to the surrealists, to Faulkner, and to their own *nouveau roman*. Perhaps they will teach us to appreciate him, as they taught us to appreciate jazz music, Edgar Poe, William Faulkner, and the American films of the studio era.

Meanwhile we must do the best we can to understand him ourselves, which involves measuring his strengths and weaknesses as a writer, and sorting out his best work from his less successful efforts. This volume is intended as a part of that sorting out. *The Owl* originally appeared in 1954, in a double volume with another short novel, also set in Italy, called *The*

Goose on the Grave. That these two works are of unequal quality has become increasingly apparent over the years. *The Goose on the Grave* suffers from a murky atmosphere, a lack of focus and coherence, as if indeed it were written by someone who felt (as Hawkes has claimed to feel) that plot, character, setting, and theme are the worst enemies of the novel. *The Owl* is altogether different. It is tightly organized. It is strong precisely in plot, character, setting, and theme. Which suggests that with Hawkes, as with many other writers, we will do well to take D. H. Lawrence's advice and not trust the teller but the tale. After its first publication, *The Owl* was reprinted in *Lunar Landscapes*, a collection of short novels and stories of which it is fair to say that *The Owl* is its only entirely successful work.

The Owl is one of the very best of Hawkes's fictions, and probably the best introduction to his work. His method has always been to work with strong images that can be developed into scenes of nightmarish power and vividness, and then to seek some means of connecting these scenes in a coherent and developmental way. Because he starts with images rather than with a story, his work *is* different from conventionally plotted fiction, though this is not the same thing as being without plot altogether. Over the years, as his work has developed, he has turned more and more to the unifying voice of a single narrator as a way of giving coherence to the events of his narrative. At the same time, his fiction, which began with an emphasis on terror, violence, and death, has moved away from those horrors toward a lush eroticism, initiated in the closing section of *Second Skin* and continued in *The Blood Oranges* and *Death, Sleep & The Traveler*. Even *Travesty*, which moves toward death, draws most of its strength from its slightly overripe eroticism—what the French, in speaking of the decadence that brings the grapes of Sauternes to their highest pitch of sweetness, call *la pourriture noble*.

The Owl is about rottenness, also, but there is no suggestion in it of what Walter Pater liked to call "a sweet and comely decadence." Only the narrator of the book finds anything sweet and comely in his world, and he is clearly a monster. He is, in fact, the epitome of fascism, at once hangman and dictator, ruling over a decaying little world with absolute and terrible authority. *The Owl* is an imaginative probe into the heart of Italian fascism, with its deep roots in Imperial Rome and the Roman Catholic church. The novel is localized and historicized, as is much of Hawkes's best work, especially his early novels. The England of *The Lime Twig* (1961), the American West of *The Beetle Leg* (1951), the Germany of *The Cannibal* (1949), and the Italy of *The Owl* (1954) are imaginative settings, to be sure, rather than documentaries of social realities. But they are attempts to reach a kind of depth, a kind of truth about human experience, which is based on historical and cultural processes. The narrator of *The Owl* is as much connected to a particular heritage as the speaker of Browning's "My Last Duchess" though he is not so precisely located in time and space. It is relevant to think of Browning here, for Hawkes has come to specialize in the extended dramatic monologue. Like Browning, he is drawn to the strange and the perverse, and he delights in immersing his readers in the voice and vision of a character whose consciousness is disturbing to "normal" sensibilities. The point of this immersion in the abhorrent is to force readers to acknowledge a kind of complicity, to admit that something in us resonates to all sorts of monstrous measures, even as we recognize and condemn the evil consciousness for what it is. As as literary strategy this requires great delicacy and control. Both the horrible complicity and the shudder of condemnation must be actively aroused by the text and maintained in a precarious balance. In *The Owl* Hawkes manages this feat as well as anywhere in his work.

The narrator, whose voice is our guide to Sasso Fetore (Tomb Stench), is calm, orotund, and self-righteous. Il Gufo (The Owl) has the title of Hangman but is also *de facto* ruler of his village kingdom. Laws were made in the past and they are not to be broken or revised. They mainly take the form of "interdicts whether cried or posted, 'Blaspheme no more. Il

Gufo'" (6). In the extremity and consistency of his ruthless complacency, Il Gufo approaches the condition of ridiculousness. He threatens to become a comic figure more than once in the course of his narrative. We read his words with a repressed giggle, a blend of abhorrence and amusement, tempered with fear. This creature is a construct, obviously, a talking fiction —grotesque, macabre, absurd. But such fabrications have stalked our real world all too frequently. Here—in fiction—it is tolerable, but it masks a reality all too like its own false face. And there is a power and an attraction in this evil. The marriageable women in the village are drawn to Il Gufo and their fathers eye him with hope. But he already has his "tall lady," the gallows, and he is forever true to her. As an unknown voice tells one of these fathers in the opening lines of the book,

> "Him?
> *Think not of him for your daughter, Signore, nor for her sister either. There will be none for him. Not him. He has taken his gallows, the noose and knot, to marry."*

The story of *The Owl* is simple and fierce. The town of Sasso Fetore has lost its young men—apparently in a war. As the fathers are reduced to thinking of the Hangman himself as the only candidate for their daughters' hands and dowries, a captured soldier is brought to the town and imprisoned in the Hangman's fortress. The prisoner, who is foreign and never speaks, becomes an object of hope for these fathers and their daughters. But the Hangman has other plans for him. Sasso Fetore is an infertile wasteland; Il Gufo is its king:

> Without mass violence, Sasso Fetore was still unmerciful, it was visible in the moonlight, purposeful as the avalanche of rock and snow. Here in the cellars and under roofs as far as the boundary, the old men slept in their stockings and the others, confident wives, warmed wedding bands in their armpits. Politically, historically, Sasso Fetore was an eternal Sabbath. (12)

INTRODUCTION ix

This lunar landscape reveals a town so gripped by law and order that it is almost dead: "As a prosecuted law with the ashes of suffering and memory carried off on the wind, Sasso Fetore was a judgment passed upon the lava, long out of date, was the more intolerant and severe" (5). Even the "cloud formations over Sasso Fetore were consistent of color, large and geometric, the clear head of a Roman heaven" (3). The Hangman rules here by law:

> To the Hangman went the souls of death's peasants, to him were bonded the lineage of a few artisans and not least the clarity of such a high place, a long firm line of rule. If there was decay, it was only in the walls falling away from proclamations hundreds of years old, still readable, still clear and binding. (2)

Though the walls may crumble, the proclamations survive and hold the living in their iron grip. The geometry of the sky, the clarity of the view, and the persistence of law shape the village and bind it to the Hangman's will. Its history began with a "primitive monastic order whose members worked in strict obedience and were the first inhabitants of the province" (30). As the Hangman says, "The immense king's evil of history lay over the territory" (40). He breathes this atmosphere and it refreshes him. He keeps a pet owl of his own, and when it attacks a young woman who had dared to disturb its privacy, implanting its claws in her scalp, "circumcising the brain," the Hangman takes it to his arm and comforts it, sending the girl home without a thought for her:

> But I wet and smoothed the feathers under the triangle of his beak with my tongue and he regained himself, once more folded into his nocturnal shape, and only the eyes did not relent. I gave him a large rat and slept near him the night. (40)

When the Hangman dreams, he sees his country as a tiny green mountain under three white flags:

The country was no larger than the flags and as perfect. The road was a bright red line winding to the three precipices and the capital of rigid existence. And the flags were moving, fluttering, the motion of life anchored safely to one place. (41)

The orderly vision of a country no larger than its flags, of life anchored safely, turns into a nightmare. The flags disintegrate into shreds and the hunchbacked village fisherman's voice comes out of the sky, concealed in the rays of the sun. He says, " 'The fish are running well, Master,' with mockery in his voice" (41). Fertility is the enemy of law and order. The fasces, ancient Roman symbol of justice borne by the Hangman's ward Pucento, symbolizes the power of the law, which is the power of punishment. When the prisoner arrives, awakening hopes of fertility and fecundation in the hearts of the village's young women and their fathers, the Hangman meets this threat to the eternal Sabbath with the full power of the law and all his authority. He turns the prisoner over to the tender mercies of the prefect, with his live coals and his four hooks of punishment, but he does not concern himself with these "temporary arrangements." He thinks of something more permanent—his gallows:

> The tall lady stood below in the court almost as if she were taking the sun. Now she had no rope. Her place had been appointed by men who with trepidation paced off the earth and tasted it upon their knees under her shadow, a shadow taken to the earth and remaining there. The scaffold itself, shaped like a tool of castigation, was constructed to support the dead weight of an ox if we came to hang oxen. She was of wood and black as a black ark, calm by nature, conceived by old men with beards and velvet caps, simple and geometric as frescoes of the creation of the world. (21–22)

The village possesses another landmark, a statue of "the Donna," a different sort of lady, who represents the softer side of the old religion—mercy as opposed to justice. Says the Hangman, "Surely the Donna made the scaffold majestic" (13). The Donna suits the "soft flesh" of the peasants, but the law gives "bony strength to the lover of Donna and legend":

> The character and the code, right upon right, crashed into the pale heart when the culprit hanged, her prayers for him so soft as hardly to be heard. She saved none—salvation not being to the purpose.... (13)

The statue of the Donna is not given a prominent place in the village. It stands before a cave in the burned forest—"an idol whose nights were spent with a few small deer and speechless animals" (27), says the Hangman. When someone reddens the statue's cheeks with blood, the Hangman is offended and simply rides past on his donkey and kicks her off her pedestal. After this, the scaffold is the only "lady" in Sasso Fetore.

If the softness of the Donna statue disturbs the Hangman, the unruly nature of sexuality is even more abhorrent to him. Indeed, when the cheeks of the Donna are reddened, he treats the statue like a painted whore rather than a vandalized icon. Later, one of the most deeply and distressingly imagined scenes in all Hawkes's fiction is based upon the tension between the bodily vigor of sex and the mental power of law. In this scene the villagers hold a pathetic remnant of a fair below the Hangman's ramparts: "The fair was pitched directly below the fortress, in good view, and for the benefit of the prisoner up there" (43). The Hangman doesn't like fairs. As he puts it,

> History had forbade the fair, a guise for flirting and the dissatisfaction of a sex—the fair invoked only when the measures of fathers failed. I listened to the festival, the ribaldry of the viola da gamba, the concert of bushes. How could it be anything but an ill omen, the distraction and the gaiety of woman preceded the fall of man. (43)

The Hangman detests women, seeing in them, as Judeo-Christian tradition has taught him to see it, the source of all human evil, which stems from Eve's breaking of the first law. He watches them coming to the fair:

> They stumbled, the swaying shanks of hair, the flaming red scarves binding torso to hips and the cantilevering, the maneuvering of the skirts. Newly shod and gowned, the purple and green of earth and sky became warm in their presence; all that was female, unnatural in congregation, came into the open air walking as geese who know the penalty awaiting the thief who catches them. (43)

This hatred of the feminine, and of the sexuality associated with it, is not a mere aberration of the Hangman's mind. Hawkes is not presenting to us an individual case study in this portrait but the deep mentality of an entire culture. The primitive monastic order that founded the community is the source of this hatred of the female and of the flesh itself. And it is the monks who initiated the grotesque ritual which dominates the fair:

> On their leaning instruments the musicians played the seldom heard 'March of the White Dog.' This whole breed had once been deprived and whipped, tied ascetically by the lay brothers on the slopes. The bitches were destroyed. And the rest, heavy of organ and never altered with the knife, day after day were beaten during the brothers' prayers, commanded to be pure unmercifully. The dogs tasted of blood given in mean measure but were not permitted the lather, the howl, the reckless male-letting of their species. Beaten across the quarters, they were taught by the monks the blind, perfectly executed gavotte. (45–46)

At the fair, the "sole remaining dog" is led in this "devilish dog's fandango" by Pucento, the Hangman's ward and lictor, bearer of the fasces itself. Boy and dog complete geometric figures, "a square, then a circle," all the while "straining to duplicate the measure, the ruthless footstep of the past" (46). In this scene, which mingles elements of the grotesque and macabre so powerfully, we find that perverse preoccupation with sexuality which characterizes all puritanisms. From the monkish thwarters of canine sexuality to the Hangman himself is only a step. In bending the poor beasts' motions to the mathematical rhythms of the gavotte and the abstract figures of square and circle the monks impose law upon nature—a law which defines itself as the opposite of nature, which becomes "law" by virtue of perverting nature. What the Hangman never tells us, of course, but what comes through his every gesture and word, is the immense and perverse pleasure this fascist takes in the stifling of what is natural and pleasurable in others. And the Hangman's ultimate pleasure is in his ultimate power. The taking of life is his consummation. The gallows is indeed his "lady."

The scene at the fair is followed by the attempt of Antonina, the most eligible young woman in the village, to win the Hangman himself as her husband. He climbs the hill of the fortress with her and she offers herself to him: "Honorable Hangman. Carino. Il Gufo. It is you I love." And this is perhaps the most terrifying thing of all: that the woman should be drawn to that very power, attracted to that very hatred of all things lawless, feminine, and natural, to offer herself to this macabre creature. What happens is grotesque, turning intensity to comedy:

> Antonina rolled stiff on the brown hilltop and the skirts loosened, lifted by the wind. She pushed her fingers into the bent grass and dragged her hair on the silt and stones. Her slender belly thrashed like all cloistered civilization among weed, root, in the wild of the crow's nest. I reached into the sheltered thighs touching this bone and that and felt for what all women carried. High and close to her person, secreted, I found Antonina's purse which she had hid there longer than seven years, that which they fastened to the girls when young. What was there more? (49)

The language reeks of sexuality here, but the deed performed is only a financial transaction or an act of theft. He takes her purse but leaves her person, touches her bones but ignores her flesh. Antonina's father tries to regard this as an act of betrothal. Il Gufo gives him no more satisfaction than he gave the daughter.

It is not my purpose here to examine the entire novel or even to offer a coherent interpretation of it. That is much better left to the individual reader in any case, who will find much of interest in scenes involving or concerning the prisoner, such as the extraordinary "judgment supper" in which is enacted the corruption of Christianity, as if its founder had not been a victim but the executioner himself. The Prisoner in this novel comes as a potential redeemer, one who might restore fertility to this tomblike and infertile world. But this salvation is not acceptable. The Hangman is in love with death, which constitutes for him the most perfect order of all. He is not a venal sadist. It is the Prefect who plies his hooks and fingers his truncheon. The Hangman is beyond this. He serves the Law and follows the book from which a "hangman knew the terms and directions, the means and methods to destroy a man" (000). He loves not pain but destruction itself. The execution of the prisoner, when it comes, is preceded by a lavish feast, altogether different from the simple meal of fish at the judgment supper—a feast so rare and stimulating that it brings back "the effulgent memory of execution, step by step, dismal, endless, powerful as a beam that transcends our indulgence on the earth, in Sasso Fetore" (62).

To the Hangman, the world itself, the whole earth is Sasso Fetore, a stinking tomb. Behind his hatred of life, of the unruly, the sexual, the natural, is a terrible fastidiousness. Execution cleanses the earth, purifies it. As he rides through his world, or contemplates it from what he calls "the absolute clarity of my vantage above" (19), he sees the men as so many "Garibaldis burning in a cold and windy piazza" (9); he sees the hunchback's daughter as "a girl who should have been burned in Sasso Fetore" (27). For whatever happens to threaten order, whether the libertarian politics of a Garibaldi or the apparent sexuality of a girl, Il Gufo has one response: purification by fire. For others, the scaffold will do. But *all* are guilty, *all* must be punished, as they were even "in the time when there were men to hang and those to spare, with clemency for neither" (44). What the Hangman hates is life itself.

This portrayal of repressive fascism is, as I have tried to suggest, at once terrible and comic, bizarre in its extremity but profoundly accurate in probing the philosophical and emotional roots of this mentality as embodied in the Hangman. It is of course an imaginative construct rather than a case study, an emblem rather than a portrait. In saying that plot, character, setting, and theme were the enemies of the novel, Hawkes was hyperbolically and provocatively protesting against certain traditional ways of approaching the construction of fiction. But the only true justification for surrealism in art is that it destroys certain surface plausibilities in order to liberate realities that are habitually concealed by habits of vision attuned only to the surface itself. And this is precisely what Hawkes accomplishes in *The Owl*. It is time to recognize that achievement.

Brown University   ROBERT SCHOLES
Providence, Rhode Island

"Him?

Think not of him for your daughter, Signore, nor for her sister either. There will be none for him. Not him. He has taken his gallows, the noose and knot, to marry."

The fathers of Sasso Fetore, their chins in hand, let me by and stuffed away the dowries, squinting at the shadow of the tall lady at my side. Though I was named Il Gufo, the owl, I was tall as she. One by one they deliberated, and well they might, for Sasso Fetore had few left who could walk on the cold stone in the morning without clutching their knees, curling their toes, bending their bare narrow backs, who could wake and anticipate the cry of guai! on the cold air. Few enviable as I. If a father had once seen Il Gufo, sucked in his breath when I was pointed out to him, he might climb again to the fortress for another look, alone. Theirs were the wobbling legs and, at the last minute rubbing tobacco under their lips and turning to flee, they had no word for the hangman. Not one had thought to put his daughter's hand into mine or expected mine to fall on hers. Now they were tempted.

But with each new day the old fathers of Sasso Fetore returned to the field, pressing here and there the suit of a Teresa, Lucia, Antonina, Ginevra. The woman Antonina accompanied her father and if his task, hasty, catholic, were not impossible, she made it easier. Even when passing the bodies of the foot soldiers, blackened chimney boys, collected in the ravines like thin logs with blackened shields or visors turned up to the weather, they devised second-best plans for making husband of the prisoner held captive in the hangman's fortress. And at the time when light rain or storm quickly darked the fortress any hour, my ward Pucento was begging for a wife.

The prisoner, delivered into altitude where there was time and silence to devour him, was the hangman's. The fortress which kept him safe was cleft in two parts on the pinnacle of the city, high tower and low tower, and from either battlement there was an iron-edged view of the world, its cliff, the tilted slopes at the bottom, the sunrise and sunset, and, not so far off, the border itself of a definite black and white. To the east it was possible to find a thin white horizon, the sea. If any in Sasso Fetore saw out there a Venetian sail, they pretended it was a dream.

The fortress was property willed to Sasso Fetore's hangman; his also the road winding from the valley, the produce of the fair set up summers at the foot of the mountain, the remains of the monastery and Campo Santo, and the barrenness of the wasted descents below covered with briar and rose campions. To the hangman went the souls of death's peasants, to him were bonded the lineage of a few artisans and not least the clarity of such a high place, a long firm line of rule. If there was decay, it was only in the walls falling away from proclamations hundreds of years old, still readable, still clear and binding.

The sun reached the mountain first and finally the valley. Each morning I waited for the light, and from the high tower, back to the steady winds, I watched the border until I saw the white skeletons of horses and men with hanging heads, the white bones of the legs sinking into the sand, the pistols swinging against pumiced hip bones. An order kept them ever moving around the mountain clock.

"Mi scusi, Hangman," the first deputy would shout, "it is time to assemble!"

Leaving, taking another good breath as I turned, my boot might kick a stone from the battlement and down it would fall until I heard it land with a breaking of shells in one of the rooks' nests that littered the golden vertical cuts of the cliff. My boots were the heaviest of all, black, laced under the knee. The birds made no sound but suffered these accidents in silence as if natural. The cloud formations over Sasso Fetore were consistent of color, large and geometric, the clear head of a Roman heaven.

By a decree dating to the Council of Bishops and Gaolers, the heart of the hangman's escutcheon burst and became an owl: with wisdom, horns, and field rodent half-destroyed, hardly visible under the talons. The bird, the scholar with his hunger clamped exactingly to the rudimental prey, peered from his shield tacit, powerful, bits of ruff and gut caught to his savage bill. The owl. And the hangmen, as they came down the inviolable line, sat or hooked the demanding winged beasts upon their necks and shoulders, lordly claws digging into lordly men, and assumed rule to the archaic slow drumming of the nocturnal thick wings against their ears, bearing instantaneously the pain of authority injected directly into their blood streams, as the owl clutched and hooted of old upon the darkened tree tops.

The owl kept watch on peasant and prefect alike. He sat erect, taking keen sanctuary among the stones of the campanile, unmoved while slitting the thin stomach of some gray animal with his beak or dusting with his low matted feathers our rubrics. On a cold throne, twisted branch, he kept a mute, tenacious, arbitrary peace, fit with little exercise, thirsting quietly for the scrap of ligament or skin needed to keep him cloaked above the city another hundred years. He had a foul breath, a deadly hold, and was quick; his eyes, and the trembling in the short armor feathers, the beak that when sated remained locked, were for the cold law and sermon that came out of the forest to make the husbandman quake. To some he brought this surety of judgment and the vision, challenging, of the broken neck. I am speaking of myself, Il Gufo.

To the hangman also went a senatorial apartment and a donkey. Faggots for the cold of winter were supplied on the hunched back of a creature, Monco, who had in his youth fallen from a parapet of the fortress and survived.

They in Sasso Fetore said that I rubbed soot on my face before sleep, that I possessed a lock of every hanged man's hair, that I confessed once a year to a Jesuit exiled from the Holy See, and that I collected about me varieties of stinging insects. None of this: my walls, ceilings, and stairwells were painted often with a white chalk, fresh and sharp as bone, and not a shadow went undetected. The sun was urged down the spires of the scaffold. But the old men of Sasso Fetore needed to talk of their hangman.

Up and down the cleared streets they went, peering overhead lest the bough break, spoke of young women in wait at home, every day combing their hair in a different style, each villa, town house, hut become an urgent convent. Most, like Antonina, were mistresses of bridal chests filled with seven years' accumulation of lace and white bodices, linen, and trinkets that were bolted from the rain. Their fathers plied for them by day, toured the countryside, down on the rocks they beat their hands under the prison windows. The girls were not merely virgin, those unseen propagated a sense of the timelessness of denial, of death hung rocking around and around on the broken-spoked wheel atop a pole. Little Ginevra was kept at home; Antonina, with compressed jaws, came to see for herself the smoke rising from the mass, the even rows of old women paddling their wash, the air of vigil and continuance over the bridled province.

"*He hangs them until dead. Will we have a race of executioners, Signore? Let her be still.*"

However, they turned to stare, to speculate on Il Gufo with or without hopes of conjugation even though the custom of ribaldry and carousing during the season of the summer fair fell more and more into disuse. The men now alive, remaining, had not often seen me in a black hood and waited for the sight.

Storms came from the prohibited east of the ocean. Green and sudden and, sometimes mixed with snow, they battered first the fortress, tearing the bartizans, dislodging and rolling into the air the rooks incapable of opening their wings. For kilometers the rose campions were pierced under the hail and the vineyards bodingly destroyed. Rain scrubbed the high cobblestones, made the straw roofs swell and doors swell, imprisoning Sasso Fetore behind the thickened jambs. Nowhere could a man walk without seeing through the rain the city's virginal design, the plan of its builders: the sheer blackness of stone intended to resist and put tooth to the howl and sluice of water, intact as it was, echoing, beset with the constant fall of the rain, unviolated and dark as in the Holy Day curfew of the year twelve hundred. No storm could dislodge those early grimly smelted chains but rather gave the city its victory in its architecture fixed, steeply pitched, weatherbeaten. Not a bolt rang. As a prosecuted law with the ashes of suffering and memory carried off on the wind, Sasso Fetore was a judgment passed upon the lava, long out of date, was the more intolerant and severe. Only the absolute wheel is known, old as it is, and I looked for the first exacting laws in the archaic, listened for the skidding of an obsolete machine on the narrow driving streets.

"He shall be dead, by means described hereafter, on the first sunless day to come, and his soul shall be said to exist no more." The census of Sasso Fetore was set with the great seal of depuration—jugglers, actors, lame soldiers, all condemned and hanged.

The snow, falling morning and night upon the head and arms of the Donna, inevitably ended in an hour brighter than any season's and one, two, but no more, children and perhaps my ward Pucento slid down the slopes on their bare haunches. Hatless, without sweaters, one or three black merrymakers scaled down the white field kicking against the snow, careful to make no noise, each figure far apart and a speck black as a devil's finger. Then Signor Barabo, Antonina's father, would climb into the town, up the narrow street with his brief under his arm and head bent as if looking for a drop of blood on the snow. He stopped to read every proclamation again. He had some divination into the past, for he knew that a hanged man's legs were bound at the ankles, that once the body is on the rope, it is at the executioner's disposal to hang it as long as he wished, a week or a day. And while thinking of this ritual, Signor Barabo thought at the same time of the ritual in which the groom and groomsmen were about to fall upon his Antonina, corset, mantilla, and all.

The road up the valley was almost impassable in the snow. A thin fox, the old man might traverse it but the way was unbroken as if the forbidden horizon of the sea had crept during the night to our walls and pickets. My donkey's hoofs split in winter, still I drove him each day under the seven-foot arches of Sasso Fetore burdened with a saddle built of wood and iron and bound to him by a girth that froze and cut the belly. The tips of the donkey's ears were cut through and a small rusted bell permanently fixed to each that jangled to his uneven gait and high station. Astride him early in the cold and short-breathed brilliant morning—the pinnacle of Sasso Fetore was like a crag summoned from the Alpine assizes—I sometimes heard the prefect's voice piping our drawn streets: "Permesso. Countrymen, good morning!" I would pull sharply on the bridle.

In Sasso Fetore all but the prefect welcomed the communication of interdicts whether cried or posted, "Blaspheme no more. Il Gufo." During the rain, or after one of the determined snowfalls, this injunction met the eye of every man and boy and was obeyed. The fox was traditionally the blasphemer, and the white length of the idle valley road was littered with the red carcass trapped in winter by a few terrified children blowing on their fingers. My ward Pucento cut the tail off close as he could. I saw Signor Barabo cover the fox hastily with snow. A dead fox meant another wedless day.

Signor Barabo forever carried with him some item intimate to the nuptial, the garter, the soldi the bride tucks against her bosom, some aphrodisiac trinket he had spirited from Antonina's seven-year store. One of the articles often in his possession was the private purse she carried under her skirts. And this he displayed under the priest's window as if a lucky impregnation might come between the secular nectary and the sacred. Signor Barabo's heart, consciousness, and ambition ended in an appendage that housed the kidney and overhung his groin like a tapir's snout—blind sack he lightly rubbed while discoursing and guardedly measuring the passers-by. He had a large flap hanging from the shoulder under cover of his coat. On winter evenings when his old wife massaged it with liniment, he struck her and, enraged, leaned over again so that the deformity rose up with uncanny liveliness to be oiled. Outside he was a peasant, inside a fish whose concealed pouches could inflate to considerable size until he groaned in his own monstrous dimensions.

"My Antonina, my daughter Antonina, will be ready to take her husband immediately after the Pentecost."

Antonina was a woman with a narrow girdle and face the soft color of an olive. It gave me the same strong feeling of satisfaction to see her as it did to spy my ward Pucento falling about in the snow, the western gradation of white marred by the far-off kicking and helpless salutes of his lengthy arms and legs, an autonomous and senseless play before the afternoon closed to the rising of the wind. It pleased me to watch him attempting to steady himself, the black straw on his head rolling across the snow field, one at least from whom the owl could snatch a river fish without trouble, so valueless was he. Idyllic his thin trunk, woman's voice, his pranks in which he exposed his shivery loins unwittingly to the power around him. Pucento as my ward remained to hurtle himself against Sasso Fetore's grave unrestored stones, a jest of flesh, angular, laughing, incompetent.

"Pucento," early in the morning I leaned over his bed, "Pucento," I said until he opened his eyes. And, but a moment awake, flat on his back and face to face with Il Gufo, he would try to answer, fall to twisting from side to side, clutching at the gray shirt he slept in and grimacing. He little thought of jesting if I caught him with the dream still on his fingertips and if I broke in upon his disarray and refuge. "Buon giorno, Pucento." As soon as I moved, he lapsed into sleep again. Waking later, he his own medium, he would bolt for the cold street shouting, "Faggots, faggots!" heels skipping high as his knee and wailing as if the wood might crawl to him at a slow-moving pace.

The parochial owl hooted and Pucento made a good deal of noise before he thought better. At night he sat in front of the wintry red brazier and I put him to work kneading my virgin rope coils. He ate cheese cut from a large white tub. The owl stood above us like a thick-chested, legless, disciplined commandant facing, between torn forests, a ragged enemy of Austerlitz.

I watched the coming of dark from the high tower. Sasso Fetore, its roofs sharply crumpled and pitted below me, grew dark but never completely invisible, always some fragment of thatch or colored glass withstanding the night. There was only part shelter in the tower; I stood with my shoulders hidden in the damp and one fist and my face raised, exposed to the rush of air, the long fall away to the villas and water wheel lost below—that curious feeling in the fortress of half human, half mildew of history, a precarious high post with the open night in front and a wet niche in the stones behind. I gripped one of the cornices. And the moon passed, small, cold, flying between the dead trees. The ordeal of older tribunals, the plagues that attended the newborn and the roof of black stoop-shouldered angels that awaited them, the fiber and the crack of the ferrule amongst the population, these I thought of in the evening when my boot heel ground under a bit of fortress rock scum.

Down, in the dark, the peasants were eating their macaroni, flat bread, and a paste made of sheep gut. On each wooden table was a liter flask; the wine was new as sap. It was as if I saw them all, corselet, benches, arms and legs, Garibaldis burning in a cold and windy piazza, so certainly were they in the outer and lower spaces, peaceful. And behind the huts, near the mouth of the cave marked by a white statue of the Donna on a pedestal, lay Sasso Fetore's forests, barbed and stripped to order. Here the female owl scratched at her blue egg with a diamond.

"*Signore. He sleeps in black sheets. Black straw lies in his stable. I would not give him my daughter—no, Master—not even if he does eat, I've heard said, the pulp of freshly crushed deer antlers. Cross yourself, Signore. Take my advice.*"

Like oxmen huddled around a milk cart in a dry rut, they had their opinions. But whether the issue that gave shape to the head and bulged the eyes was the catechism, the death of a prized boar, or a remark about little Ginevra's beauty, the old men soaked their hands in warmed oil and all their thoughts and feelings, the very grayness of hair, vanished when the hangman put on his black cape.

It was the prefect who kept the streets picked clean. I could urge my donkey to a gallop until his shoulder blades rubbed and the bells tinkled unnaturally at the tips of his long white ears, and from one end of Sasso Fetore to the other I could see dry gutters, not a dead raven or a cat, no new piling of dung, and, down below the captive sewerage. The small man, fastidious only in size and weight, scourged the lice; short, thin, his bones so many that they formed, instead of a skeleton, a mesh. Across his chest were belted two thin straps pulled to the last hole. The prefect inspected cupboard and water closet, descended into foundations hunting for ripened curd, wielding his torch and four-headed pike. His badge bore the wolves at the dugs, his brief epaulettes, his black hair hewn to the narrow side, and he obeyed, though not without pinching his little lip until it became white. He lived alone below the fortress within shouting distance of the dungeon, sleeping on a cot like a field commander. Martial drum and thin trunks of plants were hung about the walls and obscured the windows.

Of all the prefect's belongings, only his stately ganders seemed descended from Sasso Fetore's own lofty impartiality. His four red-eyed birds, though they had little to eat, covered the countryside in good regiment, their tense hungry step marching over sand, ashes, trampling the rose campions. No girl could drive these white creatures in such formation, and one behind the other, making no noise, they hunted small game and insects. To see them in single file atop the fortress —even I looked up. Or below, in the ravines, winding their way among the dead, always white, sharp-billed, not so beautiful that they would keep from scavenging. Sometimes only the necks were visible, long white necks that might be broken in a dozen places; fierceness and starvation were evident in the ganders' windpipes unreasonably stretched in God's dark and genic pillory. The fowl survived, now holding their flat honed bills high asway against the horizon. The four of them, Sasso Fetore's flock, never emitted the shrill crabbing sounds of their species but appeared on the steep slippery cobblestones in silence, checking themselves, and it was with precision, quickly, that they pecked anyone who crossed their path, thrust forward the snowy Netherlandish throats like serpents. Sometimes I tempted them with my fingers, but they did not bite and continued their marching.

Perhaps Sasso Fetore was most lofty when no one was in sight, when the owl was at his instruction and outside the ganders were poised, making little headway against the sunset, the blue wind ruffling their single file, when the rope swaying from the tall lady dangled idly like a ship cable in the middle of the voyage, and the province, sloping upward, was burnished red and gold, like a Florentine coin, before night. Faintly, in his knotty ears, the donkey's bells would jangle. And the fortress with its rocky view showed its temporary darkling life of lanterns. "Buona notte, Hangman. Good shelter," whispered the last to leave the streets. The clapper hung silent in the campanile, while it was still day, the curfew formidable over sagging bellies, over the aged coveting their anchovies in the dark, out of sight. I looked up, down, Sasso Fetore held the spirit, the law of the swift leg, stones reddening above a country that had no marsh.

At the last moment, before the windy collapse of the day —swiftly it deserted, sinking to another meridian—I myself took the road down toward the ravines and stakes of shriveling berries, keeping to the center of the way, holding firmly the donkey's jaw. A little dust rose some feet behind. And from the cliff, huddled under the balistraria far overhead, I might hear old men playing the viola da gamba, their black coats sawing musically to the left and right in the very position of skirmishers.

Many ravines were empty, offering at most a bit of crackling skin cast off from the snake, a dark spot where moisture was working up from the center of the earth; but in others sat the foot fighters who would run no more, the burned bodies listing into the sand. I rode on the edge of these pits and then through them. I counted the backs of the heads. I looked at them lying upon the earth like assassinated sheep. The remains of a water-cooled cannon were driven deep into the sand. The trenches were filled with letters, in the moonlight I looked at them, letters dropped during the summer and again the next summer on the faces of the dead. The letters were strewn upon the cinders and some had been torn open by animals. And some stuck to my boot heel. I pulled the donkey after me and could not control where he stepped.

The dead were lieges to Sasso Fetore's hangman, a chary parliament with which he met at dusk, having no voices to raise and unable to tell which limbs were lost or which ribs had been staved in the process of death's accomplishment, what weight of marrow and gut given. The numbers I stumbled on, pitifully small, did accrue in some historic calculation, out of their tangle raising an arithmetic council that gave more body to their subservience than the hair, cloth, and tissue withering. I climbed through the ashes hearing my foot turn over a trench knife or wooden shoe; and the contortions, the shrinking about the knees, the unexpected oddments I found poking in the soot, in a Roman fire, were symmetrical, ordered, suitable as the leaning masts topped by slowly turning wheels that were implanted—some signaling device, not monument—in their midst.

It was here I thought of Antonina going to her bed.

Night. I was a lonely rider who urged his donkey into an overheated aged canter on the lesser slopes, and fast even as we returned up the mountain, the road became straight and high, bleak with hoarfrost. My boots, protecting as jambeaux, rubbed his hide back near the tender joining of haunches, the spurs rang. Behind I was followed by speechless adventure, the impersonal ganders. Below, the foot soldiers must wait another inspection, thoughtlessly camped in the hollows where the summer fairs were pitched. Balcony, turret, cloister, arcade, the stones immobile in tempest, all silent and austere, the city descended a few hundred meters and stopped, formidable even to me. A city through the center of which grappled the prefect's four hooks, a place part chalet, part slaughterhouse, with scrollwork upon the gables and brass crowns upon the chimneys, a province whose wooden coffins were lined with porcelain, whose garrets were filled with ransacked portraits of dignitaries and half-eaten goose. Without mass violence, Sasso Fetore was still unmerciful, it was visible in the moonlight, purposeful as the avalanche of rock and snow. Here in the cellars and under roofs far as the boundary, the old men slept in their stockings and the others, confident wives, warmed wedding bands in their armpits. Politically, historically, Sasso Fetore was an eternal Sabbath.

Using flint, iron, and tinder, I struck a flame to the candle at the base of the white statue of the Donna to whom was attributed tolerable beauty and humility and who was thought to destroy dreams. Perhaps to her could be laid the spirit of black and white in the peasant, that suited their soft flesh in both valediction and penalty. Surely the Donna made the scaffold majestic. Week after week of Sabbaths attested to this, when the regimen set down for the citizens was not perfectly autonomous and in the blue night some worshipped object was accountable to the spirit, or the spirit for a moment was awed in some simple fashion; when the law gave bony strength to the lover of Donna and legend. The character and the code, right upon right, crashed into the pale heart when the culprit hanged, her prayers for him so soft as hardly to be heard. She saved none—salvation not being to the purpose—yet, like the virgins, stood off with low head and waited for movement from him bound on the gibbet. Her statue was placed before the cave near the forest and owl's tree.

And the prisoner: perhaps he too had his Donna or some lady to accompany him to captivity. Perhaps he too had his fragile witness in another language and blessed in that outlandish tongue. And if so, perhaps she lent him an untouchable comfort in the cell, peace, his angelo keeping him quiet as he kissed her feet or paid whatever was her homage. Perhaps she assured him of clemency and calmed him. But I would think not.

Guai! Guai!

The sound made my ward Pucento pull his hands and wince about the eyes and mouth. It was Pucento who started the rumor that I would hang him immediately after the Pentecost. How could he know the day!

Several paths led to the fortress. The approaches to the prison that appeared so inaccessible from the slopes below or from the boundaries—on some mornings the patrol raised their skeletal hands—were ready, easy, and at one turning there was even an abandoned pump as if it was here that visitors were expected to drag their mounts, here that they should naturally stop to drink, being able to climb little further. All men wanted to reach quickly, to see for themselves, either the high tower or the low, bloodying their fingers if necessary, sparing not their heart valves or horse. At a bend in the high path there were loose stones that, along with nature's debris deposited in the ruin, fell down upon the rocks. Antonina let her hair drape against the rotted pump the wood of which concealed in its grain a hundred eyes; the ganders favored the low path with their rhythmic scuttling, their search for food out of the rocky air. There was a place in the middle path, steepest and most direct way to the fortification, where from an irregular hole in the rock no larger than three fingers flowed a constant trickle of water, always to be found stirring, always seeping down, draining, from the dungeons— on the brightest of days and though it had not rained for weeks. It was water they said that cooled the ardor.

In the morning, early, the unshaved prefect climbed to the prisoner with unusual speed, as if to put to his lips the treacherous and unruly ram's horn. Bloodshot, hurrying when it was not yet his hour to come to the streets, he avoided my eye, a nightmare still fresh in his head, and as he passed he covered his eye nearest me with the glove that pinched the fingers and left the wrist bare, hardly saluting, carrying his water quickly along in a rat's sack. And already the prisoner studied the lines in his hands and picked at his incomprehensible tin insignia, the badge of those about to die in public.

And still Signor Barabo tried to speak to the prisoner. He petitioned to husband him. He asked for his name. Promising him—sure of my pardon—consigning his very procreative parts to this daughter or that and joining him to his otherwise fruitless family. And Signor Barabo was not alone. The men, oldest of all, with lids falling away from canine eyes, with clay pipes wrapped around their fingers, could be seen every day dusting, dusting the studs, the iron, the boot toes, a long and tenuous bell pull, hook-backed and hard at work polishing the war horse. They uncorked bottles of spumante. The prefect talked; and in the excitement some suppliant thought to smear the face of the Donna with blood. There was more spumante.

I, hangman of Sasso Fetore, prepared in another manner to give them a spectacle that would bring a laugh, a cry, and an upsurge even from my confessor, the Jesuit.

I washed down the scaffolding brought from the seventh hill.

I, hangman of Sasso Fetore, set in motion the proceedings that drew ledgers from their vaults and inkpots to the oak table, pushing the rafters, the spokes, the axle cold and thick with bear grease, touching these ancient parts in the law's carriage house—and through the city I had justice dragged on a hundred wheels roped to the distended body.

I went to see if they also were ready.

"*But he has a fine stand, has he not?*" The other, bowing his head, answered, "*He has, Signore.*"

And the old man, muttering to himself, would set off as if to settle the matter immediately, jogging his proposals and senility, muttering in Sasso Fetore's pantomime, smiling down the hill and moving his lips.

THE PRISONER COMES

"Permetta che le presenti mia figlia. Permetta che le presenti mia figlia Antonina."

It was the feeling in Sasso Fetore that after this introduction, she—be she Teresa or Antonina—would say, "Quando potrò riverderla?" hiding hair and throat in a tight veil. During the warm weather they stood out on the rocks to watch, a young woman standing white as the Donna upon each stone below the cliff. The romantic gambados of all such as Signor Barabo—these aging courtiers—tempered the season, and the cockle cap, the red noses, turned to home greatly astir. The devil sat between Signor Barabo's shoulders and breathed heat down his neck. There was an air about Sasso Fetore—it was felt in the council, near the pump, or behind a stained glass—that one wretched mass of the sex was about to rise to the temptation.

What one of them did not expect to husband my prisoner?

Signora Barabo, with fiery arms, bathed her daughters in an alcove behind the farmhouse and the water ran from under the bristling swine, seeping across the yard from the bath: first little Ginevra and then Antonina, sparing neither, the shouts of the three women loud as slaves naked in their stockade. And not a word from Antonina when, accompanying her father, she carried pressed to her bosom a book from whose foot hung a flower, the Spinster's Needle, hardly moving under the clasp of her hand. In Sasso Fetore female fingers were powdered and ringed, were driven carelessly into dough or the common crush of blood apples, or took the axe to the pig.

It was at the time of the bath one day, when the owl also spit and picked his front, that my ward Pucento led a small column of soldiers, and amid them the prisoner, up the hangman's road. Pucento, the lictor, was well ahead and he was covered with the camouflage of torn coat, berretta, and dusty face. Pucento led his band past the girls' shifts drying on the barbed wire and forward to the steeper grade at the same moment Signora Barabo splashed Antonina from the bucket.

Pucento brandished the Roman fasces over his head. The bundle of straw and the scythe blade, arms of authority, these he kept thrusting left to right at the open road so that the rushes whistled and the blade hooked into the air. It was an instrument he might shake at the path of a fiend or rattle on the day of excoriation. Every few feet Pucento, still on his toes, suddenly crashed the jacketed axe head down upon the earth savagely as he could. I felt also his extravagance and I too the compulsion in his various steps. And as this small figure flexed his legs, his arms, even from the tower I could see that he was capable of cutting down the bush and huts as he passed.

The marchers that he led huddled together and tramped up the hill. They had been joined by one, two, members of that outermost patrol whose skulls were cavernous and whose muskets were covered with the bleach of bare collarbones. They walked heavily and clung to earth, often making a cordon out of their arms, knapsacks and blackened eyes; their greatcoats billowed and exposed the home-sewn sleeping shirts grained with lice. A rifle raised, I saw the black expulsion of shot and ball signaling their arrival.

They had come far. Still, close as they were to the end of the march, they crowded together, menacing their prisoner as if they could not bear to give him up.

A small child ran from a sty and threw grit into the midst of spiked boots, perspiration, and white lips. It was not long before those few figures old and bent out in the fields—accustomed as they were to till from light to dark without looking up from the furrow, so alike and so menial they could not be picked out from the clay and pinch of seed they attempted to sow—dropped their rakes. From several corners they hurried toward the road to stare after the band smelling of the capture. A foreign crow glided above them and watched for refuse. An old man shouted: "Come si chiama? Come si chiama?" excitedly, squeezing his hands. But if the prisoner understood, he dared not answer.

The wind tore to shreds the clouds about me, the taste of morning coffee and cognac was on my tongue, and so high on the fortress walk one could hardly imagine the burnished air and the moisture that lay in the beds of the bullocks at the foot of the valley. The rooks cowered in their nests. A spirit bottle had smashed on the stones in this deserted niche of the fortress and back and forth my boots ground the glass. It was a cold vigilant place. As I watched the progress of Pucento and his patrol, my mind measured off the fields like parallel rules: a hundred kilos would be reaped from that one, another hundred from that, in the middle field there was nothing except the grave of a peasant's small boy. The earth looked like the mud holes of rice flats, it stretched away only to provide a surface in which to hide excrement. On the eastern horizon, on the sea, a lone sail leaned down skirting the salt caps.

Then Pucento was nearer. They approached the gate that was proud only of its immunity after a past of doges, conservatories, learning and brotheldom—the Renaissance driven from our garrets and streets. The foot soldiers were close, the green skullcaps, the rusted saber, the empty sacks. They disappeared into the arches and piazzas and the smell of burning twigs.

I was spared the sight of the women holding their bosoms to the window in welcome. Sasso Fetore gave them its unlatching of shutters, and they were followed by an old woman offering them a hot pullet. As the soldiers passed the coffin maker's shop, the old fellow poured a drop of spumante into the tin cup of each and tried to detain them.

"Where is your daughter, coffin maker?" growled one, taking his draught and flattening his boots on the stone.

"Inside, inside," was the quick answer. And, "Lucia! Come down, daughter!"

But they were gone and he drank a bit himself and sat upon the curb. Two children, dressed only in shirts, ran ahead and pounded on the doors, and their running feet sounded like a monkey clapping his hands. "La, la, la, la," were the exclamations within.

They went through the streets barely catching the pieces of bread and meat thrust at them, and they turned the corners all at once like a band no longer familiar with avenues and pedestrian courtesies, took large strides, men just come from marching in the snow. They were conscious of the weapons, the powder bags hanging to their bodies. They were poorly shaved, having cut at their faces out of doors and in cold water. But even those in Sasso Fetore had gray cheeks and oily hair, the marks of their Latin, circumspect temperament. Even Signor Barabo had inherited his weakness.

And men as well as women appeared at the windows to wave stockings or a wooden fork, wiping the wine from their lips and sleep from their eyes. "Che cosa, cosa?" whispered the very young until they were lifted to see. Pucento led his men upward through Sasso Fetore.

From the roofs below me rose a crematorial smoke which they had fanned to fire in the middle of their stone floors early that morning. The first awake had shaken the second in his blanket and then the third; with the confusion of dawn still upon them, they lifted dirty heads from their arms and a vocabulary humble as the sounds of animals was put to use again. All of them wanted to touch the prisoner who had not a moment before been brought from the fields and ditches into the streets and bolted structures of Sasso Fetore, past the sagging figure of Signor Barabo himself gripping aloft a handkerchief and shouting: "Viva il prigioniero!"

From the absolute clarity of my vantage above—the wind was strong enough to prickle one's skin but leave the eyes starting from the head—I surveyed the slate roofs and beneath them those who hurried in anticipation through the morning.

The band emerged from the edge of the city and faltered a moment upon the plateau of the last back houses and over-toppled fountain, shouldering their arms before crossing the graveyard from which started the paths to the fortress. The eager crowd did not follow them from under the pitch of roofs and balconies, but the women and children went indoors again. Two old men tipped their heads together, held their arms up in excited tremor-boned salute, and went their separate ways, listening with their ears to the lock holes for what was to come.

Pucento thrust high the fasces and twisted them around and around in helpless signal, unable to move further, paralyzed with wine. The others pulled their coat collars over their mouths. But even paralyzed, Pucento's standstill was tormented with the sound of dogs leaping with torches in their jaws, his indecision comic and fearful as a clown's.

I leaned over the parapet. "Avanti!" I shouted, and at the command they rattled forward, striking their pennant staffs into the ruts. The owl hooted. The climbers in their bulky clothes tried to balance by clutching the blue and black scrub trees that twisted horizontally from the cliff. The hidden villagers sent up a low-lying olive smoke. Then I saw the prefect come out, finish his piece of dried pepper, hastily tuck his shirt into his trousers, and descend to meet the prisoner. He was short and wiry in the light. On his belt he carried a ring of keys and in his hand a pair of wrought-iron pincers. He poked his sharp tongue against his cheek thoughtfully.

The ganders cut between the prefect and the prisoner. The birds, dragoons of the farmyard, precariously trod the sheer ledge, and continued unswervingly, keeping formation in unlikely crags and watches, pointing their sails, languid but regimented and stately despite webbed feet. White and silent was their propulsion. Between the prefect and Pucento, the ganders sailed with prohibitive bodies the shape of vindictive Flemish women's headgear wide and crackling beyond their backs.

I stood tall, shoulders hard against the round of the high tower, surveying the radii of earth not quite a maremma. And far, far off some low women beat their loaves upon an ox's haunch. All about was the deceptive blue sky. In, out, to the chest and extended, I breathed deeply of the air the wind hath frozen in his belly. I hardly noticed the black tiles on the roofs directly below and did not know whether all were impassioned of the prisoner thinking he would escape the scaffold or, more likely, thinking I would hang him. But I did not concern myself with the whispering women or with the deceptive plans of Signor Barabo. Nor with the prefect's temporary arrangements for the captive, for he existed only for this, the torture that demanded no strength from his own arms, the turn of ratchets, and a blowing of bellows into live coals, his activity confined to the adjusting of his four hooks —a cunning that grew in the dark, that filled his lungs with the smut of burning hair and oil.

But I would sooner my boots pick up fresh dung by the hour, the rain splatter my bald skull, the ice stiffen my red cape, and the dwarf trees lash my shoulders at a run, than sit bent over, afraid to injure my windpipe, studying the silent drops of moisture on the aquiline nose of one being garroted in a cellar thick as a furnace. Signor Barabo could recite the canticles: 'the hangman shall be four cubits tall or more, shall have a head of prominent bones and smooth on the top so that all admire the irrevocable round of the bone and largeness of brain, and he shall be bareheaded except on each day preceding the Sabbath named in advance by the hangman when he shall wear a pointed black cap the better to see him and to make the grimness of his nature apparent by contrast with the conical black peak, in the manner in which the fiercest animals are fashioned with some unnatural largeness of hindquarters, stripes on their sides, or horn.'

Between the high tower and the low was a rampart of moss and stone, a catwalk connecting the two as stolid as a Roman battlemound. It led one at a level with the chimneys and unearthly shrubs. There was no handrail, here the rock was soft and sloped toward the drop to the valley, narrow and red with the iron secreted in the crevices. I walked out upon this ledge, the wind swept it with his frost. As I moved, a large barren nest, insulated with old feathers, was suddenly flung into the air where it hovered a moment, a round tangle of black briar, and, striking an invisible pit, fell down heavy as a demon's iron halo.

The low tower barely held one man. I was able to look over its edge into the silence below. It was a mere turret fashioned from the silence of the fortress, protected from the wind, and covered with drops of moisture like the hollow of a tree trunk. To wait here and to watch, confined in the lookout overhanging the quiet court, made my crossed arms patient and the eyes dilate as through the gray of the dark. The stone that came to my chest hampered even the intake of breath, yet at that moment I might have already fallen, stepped into the beating air.

The tall lady stood below in the court almost as if she were taking the sun. Now she had no rope. Her place had been appointed by men who with trepidation paced off the earth and tasted it upon their knees under her shadow, a shadow taken to the earth and remaining there. The scaffold itself, shaped like a tool of castigation, was constructed to support the dead weight of an ox if we came to hang oxen. She was of wood and black as a black ark, calm by nature, conceived by old men with beards and velvet caps, simple and geometric as frescoes of the creation of the world. She offered no retraction or leniency once death was in motion. Much about Sasso Fetore was told in the idleness of the gallows. And idle, without her tongue, how stolid, permanent, and quick she was, still ready as the roots of ancient speech for the outcry. Her shadow changed sides while I watched and it became more cold. The sun did not come too close to her.

"Grace to you, grace to you, Hangman," shouted the citizens of Sasso Fetore and put flame to their torches though it was still light. The faces in the crowd that formed on the swept cobblestones when the prisoner was hurried into the fortress, the faces whose noses, cheeks, foreheads—of a greedy man, sullen man, obedient man, choleric serf—struggled to free themselves from the caricature putty and paint of their daily look. There was a murmur, the press of the people, in expectation.

If Signor Barabo had come upon me then, he would not have bothered with his long practiced and formal 'Permetta che le presenti mia figlia' nor bothered to bow or hide the worn patch in his official cummerbund—he was flushed and might only have shaken his head at the obvious color and fulsomeness of his daughter's arm. But he remained now behind closed doors in the caffè and the glass before him was kept brimming.

"My eldest daughter—gentlemen, Antonina—has become the belle. Let us be bold: the word is *appassionato*. Already, sirs, I seek to hire two backs at least to carry the dowry, two more her gowns and the rest of it.

"Let us call her 'respected bride.' Donna, but we can agree, every bone in her body is true, the knuckles are hard, and the heart does not flutter needlessly. Put your coats down, gentlemen, for my daughter.

"She has been standing these seven years—confident the while—in a pasture of chilly reeds and not once has the wind spied her ankles. Have you seen them, gentlemen, the virgins touching each other's cheeks by the mill or the lonely south crèche?

"But, gentlemen, even the little girls titter when my daughter passes. The children suspect something, my friends! Will it be the white gown or the gown the color of pearl? The embroidered linen or the plain? Shall we order radiant dignity? Will we serve pheasant and fish or goose? What sort of rings? I ask you. I ask you to think of Antonina. . . ."

"There is Lucia, also, Signore."

". . . Yes. And of Lucia also if I may ask it. Think, gentlemen, for your pleasure and make your decisions. The baking must begin at once. For I tell you, the weddings will sweep us like grass fires. . . ."

Signor Barabo rubbed his hands as if he were rolling a great ball of unleavened dough, the fathers of Sasso Fetore raised the bowls to their lips in salute. A small lizard ran across the floor and the coffin maker reached out his boot and destroyed it against the beam. There were two narrow windows in the caffè, each filled with small thick cubes of purple glass sealed by lead into the frame, through which, thick, opaque, the street and the world beyond were dimmed.

And out in the street thronged the crowd. A few, grinning uncontrollably, rapped on the glass, an aimless communication of the drift of their mood. The emotion went among them, possessing their hands and feet, making them noisy. Those who had been asleep asked, "When did he come?" "Citizen, only one of them?" "When will we see him again?" Some, with uncomprehending eyes and jaws slack, shouted, "The hangman has taken his wife, he has taken his wife!" knowing not what they said.

They put their hands upon Antonina, crying, "Fortunate, fortunate, she here will not have her purse for long. . . ." The women dropped combs from their hair. Now two, now three, fell from the crowd and stood looking at each other with wonder. Signor Barabo's wellborn eldest daughter, Antonina, was no more able than they to resist the street crowds. She walked in the direction of the Jesuit's chapel, but she did not enter it. She was reluctant, yet Antonina, also, was not to be denied and climbed to the heights of Sasso Fetore, as women will when they have heard of a disemboweling or other fascination. She did not shudder seeing her sisters with the cloth off their shoulders.

Antonina was one of the virgins who have grown knowledgeable as an old wife, select at nursing fires or calling back the dead with circles and chalk. Up she came through the rose campions and sourweed, having refused the accompaniment of little Ginevra, and the tears of the brisk wind were impersonal upon her cheeks still olive with the pigment of generations.

"Antonina, Antonina, this way," hailed Signor Barabo from the door of the caffè and waved the handkerchief he had lately shaken at the prisoner. His daughter gave him the rare crow's wing of her smile. "Gentlemen," shouted the father, "behold!" And they gathered behind him in the door frame, a dozen pairs of eyes and a dozen goblets.

In a curious manner, with intoxicated bravado and a fear lest the words not be said at all, Signor Barabo shouted his daughter's banns loudly, each time incomplete: "Antonina, respected and loved of father and mother, has become betrothed by law and by permission of her father, soul, spirit, and good temperament to Signore . . ." "Gather the white flowers, gather the white flowers, Antonina shall marry . . ." "It is declared, this woman's betrothal has been fixed on earth and I, the father, give her, I give her . . ." And these speeches —he was clutched by his compatriots in the doorway—were addressed to the oglers nearest him, the red noses and scar cheeks of the women who replied by opening their mouths as if hearing a proclamation of punishment instead of the banns. Then he freed himself and pushed a way toward his daughter. "Will of the Donna that she marry, will of the Donna that she enter the rooms of the tall booted hunter; no daughter shall refuse to give herself and her father asks it. All the public shall see her enter the stranger's house, none see her return. Begone to wedlock, then!"

He took her arm and stepped forward, bowing to the left and right, low, and the integument wriggled under the coat between his shoulders. Antonina was a head taller.

"Un momento, per favore," muttered the coffin maker, catching up with them and pulling after him his daughter Lucia. The two pairs climbed the black and white streets. Black rings and bolts, white stones once washed by the sanctimonious noviates, how the wind whistled in them, preparatory to thundering down the mountain. The two old men held tight their virgins. Antonina, despite disbelief and the black band fastened about her throat, walked sternly a little in front of her father, hastening to see for herself. Her face was like the Donna's. Already she knew what propriety was lost—but Antonina's heart made public would perhaps put them to silence. Her mother, deftly, had pinned a small male figure of silver to her hip and there it caught the sun.

The red and the gold were gone from the fortress. Signor Barabo took his daughter in Pucento's late footsteps, grimacing as the shrubs slipped through his hands and he tottered, wiping his face. The fortress was black and white with age, the rooks screamed the cry of a dying species.

Signor Barabo stared up at the grated window of the fortress and as soon as he got near, he bellowed: "My son-in-law, future son-in-law, look here!" stamping his foot and shaking his black hat in the evening, unconscious of the shadow at his side.

I, on the other hand, took my usual ride that evening. Despite the disturbance among the people, the graveyard was still and softly grained as an etching. Since some had worked this day, I passed the trees still smoldering, the stunted pines whose roots they labored to destroy by burning. I rode across their farms and through the middle of Signor Barabo's villa as well. There were no signs of the young women waiting out upon the rocks. Perhaps their mothers were plying them with mulled wine. Certainly they were not teaching them any matter of weights and measures—even the women of Sasso Fetore were acute in the practice of weighing sin in their hands like a pound of oats—the sciences of law and balance being this twilight abandoned for the plaits of the true lover's knot.

I did not stop. My donkey trotted jangling past the very light from their windows, so close his hoofs shook the earth of their floors, but there was not a question of my dismounting, not a movement to detain me though the donkey brushed the sides of their gates. One would hardly know what they were thinking, except that I saw a creature leaning over a hooded well and, through the dark, laughed at her reflection in the green water. And in front of a hut there fluttered a fresh veil on a tilting weather-beaten post. The women of good station and poor were distracted from the bedrails, the pitchers, the pillows grown small and brittle under their heads. Large women such as Antonina now fitted their bodices in the moonlight.

The farther I rode from the city of Sasso Fetore, the better was the view of the fortress, large and silhouetted, black and irregular up there as the multi-prisoned bastile town of Granada. What fate had it to offer the husband hunters! Or the fathers also who, had they the stock, would have busily slaughtered the hogs and cut from them the delicacies of the viscera to be served black and steaming as roasted chestnuts.

To turn one's eye from the immediate rock, the cannibal briar crouched under cover of some phosphorescent leafy plant, from the immediate valley road crooked and hard to the rider, to twist about in the saddle and back there, high, far away, see the white fortress and its stripes and shadows, black with its secrets, terraces, barred apertures: the landmark forever there, from which formerly flew pennants of the 'festa,' now sporting crows' nests and rooks' nests over the smell of dead lions in the arena.

The forest was full of activity; on all the trees the small twigs were newly broken and some trunks were still aglow with sparks. The donkey took me heavily through the underbrush, not starting at the fired tree trunks but shying testily when we neared the mouth of the cave. The tunnel, polluted during the time of the fairs, was now lit by torches, pitch sticks wedged deep in the rocks. They had picked the forest bare, dancing a quadrille almost extinct in Sasso Fetore.

The Donna stood before the cave, an idol whose nights were spent with a few small deer and speechless animals. The donkey turned a sharp foot, the lumbering saddle creaked, and even this far I saw the defamation, the Donna's face smeared with blood. I galloped. As I passed her I raised my boot and, ramming her chest, dislodged her so that she fell and rolled upon the crackle of the clearing. And there came the peculiar rush and windmilling, the sound of a bird striking in the dark the outermost protective net of leaves, as the owl first beat his wings attempting to penetrate to his roost at dusk.

THE SYNOD AND THE SENTENCE

"Mi scusi, Hangman," cried the deputy, "it is time to assemble!"

The hunchback went first with his load of faggots and after him, one by one, the council entered, folding their way into the dark, the draft of the senatorial chamber. The old men took their places at the table and awaited the burning of the fire which was preceded by a cold unsuccessful time of acrid smoke. The hunchback kneeled and blew, twisting the faggots with a bare hand, poking his tongs.

"You too then, Hunchback, you have a daughter." Monco peered up and grinned at me, splintering a twig with his knife, and I at him, for Teresa was a girl who should have been burned in Sasso Fetore. He tore a bit of rotted cloth from his sleeve and fed it to the sparks; nothing burned so well for the firekeeper as what he wore. But it was a slow fire and there was whispering at the table.

"A man with a daughter, your honor—Grace to you for having none—cannot think of cutting wood!" grinned he again, lifting his lamed back like a tortoise.

There were twelve seated at the table, far apart and separated by such broad wood that now and then they put their bellies upon it, stretching the whispers. About their throats they had tied the traditional magisterial ruffs, collars like honeycombs of rag paper. The uncomfortable ruffs scratched as they talked, louder and louder with the hurry of the words. Each had a gavel fashioned of rolled calf's skin.

I took my place at the head under the tester blazoned with the escutcheon of a burst fess, a black leathern tapestry from which the dust was never blown, reputed to have once been the skirt of a barbarian conqueror. To conquer without going to field, this spirit of mettle was vested with predaciousness in the hangman.

"Grace to you, grace, grace," said each as I sat and the first deputy posted himself at the door, carrying athwart his broad sword. The fire sprang up and the hunchback and my ward—Pucento was weary and laughed the more—both on their knees pushed and knocked the live coals across the flagging.

"How does the prisoner sleep this first night, is he cared for?" asked Signor Barabo. Before the question was done, Pucento shouted, "Hooks!" and put his forehead down to the stone like a frothing jester. The reeds from the fasces were still bound about his thighs and ankles and some, catching a spark from the embers, smoked, shriveled to black char.

The judgment supper was served, a formality that appeased the instincts of the council, food thought to bind the officials to the hangman and to perpetuate the feast of the law body which preceded Sasso Fetore's original compulsory execution. Before each place was set a platter containing a fish, as long as a hand and thin, speckled green and served whole with the tail and head gray-black in color, bright and opal. I cut my portion with a three-tined fork and a dull knife, but it was customary for the councilmen to take the fish, and its slight dressing of oil, up in their fingers. The river diet was never changed, each day the hunchback wedged his sieve traps into the rocks of a vapid waterfall and awaited his catch, relishing most the moment when it was time to pile the fish together on a peeled stick, thrust the point through their blinking gills.

The thirteen waxen fish, lying straight on the porcelain platters, were devoured again; the councilmen once more broke the tough skin and immobile, formidable under their broad brims, separated the white from the myriad hair-sharp bones. The fish was bitter, its meat came apart with the elasticity of muscle instead of flakes. Occasionally there was a portion of small black roe. This was the cuisine of justice, by firelight the minute fish and the pointed fins were illumined, sticky, each one silver and half-picked against the great black of the benches and under the gray hands of the councilmen. The fish was the fare of all the verdicts delivered and with its complicity of bone and deathly metallic flesh, it had the character of a set jaw and seal ring. It was by the fish that the jurists earned their title Mongers. The season of salting the furrows with the fresh water's spawn was the season of many bells when the Mongers brooded upon those to die.

A food of no sweetness, small pleasure, but the same distributed at the council of Bishops and Gaolers, the same that stuck painfully in the throat of the past—Signor Barabo ate his share quickly enough. He was heard to mutter, "Bella, bella," between mouthfuls though the meal was declared a period for silence. He was tired of seeing each morning begin with the scrubbing of his daughters, was glad now his wife might put aside the wet bundle, her broom. Signor Barabo was volatile and held the fish up by its tail, the shadows taking dark hold of the ragged cavalier cut to his uniform.

I rapped my knife on the table and passed it twice sharply in front of my face, blade toward the lips. Signor Barabo, fat as he was, pugnacious as he was, kept himself quiet after that. The twelve old men concentrated until the last of the fish was gone. The hunchback went down the left side then the right, taking the plate from each as he was supposed—

assuming the familiarity that might be expected from the serfs with their rock plows or from the wine pressers—tapped each councilman on the shoulder so that the citizen should be represented and their fraternal feelings, like the tipping of a hat or the offering of swine for mercy, should be expressed. But he did not touch the hangman. When he took my plate, setting down the tall stack of the rest, he pulled a crust from his pocket and rubbed it in the remnants of the fish oil. Then ate the crust.

The moon appeared beyond the high windows criss-crossed with iron. The moon, having suffered in the heavens some voracious attack by night-migratory flocks, its face having been picked by the wind, drifted low past us now in shreds of yellow against the darkness, and disseminated the cold of its center over the roof tops, the priory buildings of the fortress. Inside, the fire took some of the chill from the room. Yet it was cold with the damp smell of their boots, beards, brown hands, and with the machinations within the councilmen's heads, each one thinking his own daughter the kindest of tongue, the best of proportion, and the keenest to do marital bidding.

The old men were dressed alike. The men of Sasso Fetore, and the Mongers with them, at a distance could not be told one from the other except by shape or peculiarity of walk, and these marks too were not obvious when they were gathered in crowds. The brown shirt and the wide-brimmed black hat covered all of them. There was hardly a variation, earthen cloth for the back and a headgear adopted from the primitive monastic order whose members worked in strict obedience and were the first inhabitants of the province. So the shirts were still the brown color of the lay members' robes and the black hats unmistakably derived from those creatures who chanted while fighting on the early slopes of Sasso Fetore, pair by pair beating each other with hard fists under the watch of Superiors—these same fighters who during the matin scrubbed the garments of the sick and diseased on the rocks by the river where the councilmen's fish swam.

Despite Signor Barabo's sash, the twelve Mongers sat before me in such lowness and humility, the fattest and oldest for all their years and their daughters still covered only by the cloth brown as the furrows and dead grapes, the brown that flooded far as the black and white posts of the border. The hunchback's fire and the cold-smoked walls of the chamber allowed no other uniform for the tribunal: by common dress the disparity of their height and features should be shown trivial and in no way belying the meanness of men.

The first man to rise, rip open his hempen shirt, and expose his breast was allowed to speak, the rest thereafter not needing to perform this ritual. That night it was Signor Barabo who could not tear at the buttons quickly enough, scratching the white skin in his haste. The others, who remained seated, buried his first words with the crash of their gavels. Antonina's father revealed greater and greater portions of his chest, appealing to that which he was not and could not hope to be.

"Hangman. *Boia savio*. I asked if he was well cared for," he began. "There was no answer. I sought to learn if he slept and was not gratified. And I climbed to the fortress, *Boia*. Yes, straight up without admiring the view and risking myself to the steepest path. I called "Son-in-law!" using that familiar term. His face was easily seen between the bars and brambles but even from him there was no answer. It is boding. What has been done to him already? Are we not to welcome him when the sun warms us?

"Look at me. I show my chest and part of the belly also, and I ask: *Boia savio*, what is to be done with him? Are the dead to be made jealous for nought? Who would make Antonina—think for a moment that she is not mine—who would make Antonina shudder again? Must our young women walk with their faces forward and forever scanning as if they be the figures with salty bosoms leading ships by iron noses into the gales? *Boia savio*. Not she.

"Should not my son-in-law walk freely? What have I but to take his fecundating germs like pieces of eight for my daughter? And to let him build some cradle, gentleman's cradle, if he can, out of the deep. I've thought of it, so I might lean myself between the two of them when my old wife is dead, and hang to the arm of each when they take me airing. I'd hurry my decrepitude for that. No one else, Hangman, has come or here offered himself as yet. . . .

"And yet my knuckle, with the rest, has surely tapped their breastbones, and in the past we've hung them. Rather you have hung them, *Boia*. What will our justice be? If it is a white card—a white card, *Boia savio*, what mercy!—think how my son-in-law will sleep with his snoring and rise early to stretch his white legs, while I sleep late! Think how Antonina will lift her eyes! Who knows what his inheritance might be.

"I would have it so she will her husband, one to whip the hostlers when the hostlers return, and be regenerative after she has collected a thousand lire of his passion.

"That. Or hang him high by the law?

"Which?"

I gave him my answer. He shook once and attempted to button his shirt. It was as if I had ordered his soul bled from the arm, and Signor Barabo sat down. It was impossible for the old man not to betray himself, his disappointment physical and evident as the rings round the owl's eyes. He looked at his fingers, thinking vaguely there should have been large blood-gems set upon them. I gave him my answer again, adjuring him, and the Mongers listened on either side of him with faces like millstones, turned to profile and blinking. There was the coffin whose face was nailed together, the cart with his elbows resting upon his wheels, the wealth of land and his earth belly, and Signor Barabo who was not loath in his heart to see the prisoner die. This oligarchy—the mud of their thighs no better or surer to last than that of the old man who left his rake to watch the prisoner pass—was yet familiar enough with the ancient tongue to understand me, these old masters having in their histories sentenced not a few.

High over our heads there was a rustling, a tearing as of a sailsheet, and from a nest suspended by the beams dropped a half-stripped bone to clatter upon the table. Signor Barabo

THE OWL 33

asked to excuse himself. Monco the hunchback burned the last of the fish in a heap on his tall fire. The air of longevity was strong, the Mongers sat straight as those before them, it was a hall in which the lizards in summer, the rats in winter, peered at the justice of Sasso Fetore captured in oil paint, the noses and downturned months preserved by the sculpting twist of the palette knife. The building had a flat roof and was surrounded with arcade and flagstaffs and marred by the slings of barbarians. It was the only white building in Sasso Fetore. Here, this night, we conducted our business despite the look on Signor Barabo's face and while he trod through his dreams keeping one hand on the pocket where he carried a curl of Antonina's hair.

How long shall be the length of rope? That was decided.

And the braid of the rope, fine or coarse? That was decided.

Was blood to be drawn first from the throat or not? That was decided.

And who shall be witness? Also decided.

At last Signor Barabo began to wake and to his brothers' voices added his own yea or nay. Still he interrupted to murmur, 'The respected bride, the respected bride,' and sharpen his eyes as if to contemplate the beneficence of this title, yet envisaging a wedding summons of four quaint pages.

The firelight seemed to come out of its hiding, their faces became white and more round. There was a cut to their lips and a leathern round to the knees beneath the table. The hunchback—the rock that bowed him rose on his spine higher than the head—waited upon each old man with an iron pot and from it doled half a hand of ash on the wood itself in front of each; to each he gave a quill and poured water over the ash, the writing fluid thick and black and staining the wood. They commenced to scratch and blot the document as it passed up and down their row.

"*Attenzione*," they again would read along with the rest from the northern wall, "*Attenzione* . . ." I could see the rain beating the shoulders of those who depended upon nothing so much as the formality, stringency, the trueness of ring, the evidence of the Death Decree. "*Attenzione!*" Always the Decree contained the name of the condemned. It was the task of the citizenry on hands and knees to scrub the piazza after the execution.

"*Boio savio*. Antonina would wear twelve skirts to her wedding." And he wrote his full illegible name in the script that once flowered in this country. But Signor Barabo was also a cruel man, as his daughters told.

"My daughter, Altezza," grimaced the hunchback, "would wear none!"

My face showed nothing. Like long trained and impounded clerks, the old men dropped their quills and took up the gavels: "Grace. Grace to thee, Gufo." And they followed me into the night, leaving the faggot-gatherer to quiet his flames. The Mongers huddled together and waited to see me depart before dispersing, before touching each other's shoulders silently and descending alone to their several rooms, knocking secretly on the thick doors in narrow streets. Think of them then finding their way about among their personal belongings, correctly choosing the bed, the roughened walls, and those who have kept awake for them. How unsearchable is the law whose sentence they subscribe to and which leads each home and to a sleep that continues while the chimney cools! We left them, Pucento and I making a hard noise up the slope.

We went the way from the senatorial chamber to the square of the tall lady, past the hangman's leaning house and stable, and up the highest rampart broad enough to permit a cart of fifteen unfortunates to travel—before dawn and with hands roped—to the gallows. A few posts like black briar still remained driven into the rampart; from these Pucento hung out, peering, a bare-headed silhouette, toward the nightly star-spaced distance that drifted over the whole of the valley.

"Padrone," my ward pointed and whispered, "the prisoner."

Up there was a window and two thick bars, a window secreted in the crevices of the fortress, an opening eaten through the stone so small as to hide the features of the prisoner and stop his cries from passing to the air.

The portcullis was only an arm's length over the prefect's head. There he stood idly, a truncheon dangling from his wrist, and smoking a short butt of cigarette.

"Buona sera," he whispered and made a gesture to straighten the two straps snarled across his breast. "A respite, *Boia*," he said softly and shook the fingers of his truncheon hand as if they were stiff and pained him. He had oiled the keys and they glistened at his belt, those that opened the locks of coffins as well as the fortress door. He smoked and the thin substance of his cheeks stretched over the bones to the mouth. He shifted, the rotted pumps on his feet scratching the ice and gravel, and he glanced between the teeth of the portcullis toward the clouds.

"Will it storm, Hangman?"

"Prefect, I will see my prisoner."

Still he put no life into his marionette arms and legs, thin akimbo creature resting, inclining, bedraggled in his official position. Then he began to study his keys as if he did not know them well. Sullenly he thrust them toward me.

"Can you choose, Hangman?"

I chose the large bull-headed key. Yet he did not move. To the west stretched the topography of our lowlands—the snow was collected in the pits and gullies and with the flat white shape of salt encrustations gave tooth and rib to the night, dimly providing a body for the dark. The low possessions were there, wind-swept, across which barked Pucento's fox. The prefect's name would never be scratched on the floor of the senatorial chamber, nor would he have so much as a tombstone to slant abjectly through the lean centuries to come. He existed as at the mouth of the drainpipe to Sasso Fetore. But the prisoner was, for the moment, in his custody.

"I will need a lantern, Prefect."

"Carnefice. Eater-of-deer-horn. Will he not have a term of servitude? I should care for him, Hangman, as he deserves."

The prefect made this plaint and leaned more heavily, refastened the keys to the leathern girdle beneath his tunic. His kepi, with its battered top, was ill-fitting and crooked; it slipped far forward, giving outlandish shadow to the shape of his head. There was no sword in the scabbard which he still dutifully wore where most men in Sasso Fetore would carry a good hip.

"If he has not escaped, Prefetto, I will look at him."

The bolts fell. Pucento began a soft excited keening as we entered and, before we reached the dungeon, was calling ahead *prigioniero! prigiomero!* The prefect faltered with the lantern. I walked slowly, my footsteps paced. But Pucento ran suddenly, flung his body against the bars of the cell, and shouted loud as he could: "If we free you! If we free you, Mostro, will you not ask for the hand of little Ginevra!" And Pucento panted, the straw on his wrists and ankles rushed against the wooden bars.

"Look now, Hangman," the prefect swung the light at arm's length, "and decide, per favore, to leave him with me."

For now the prefect was the proprietor, knowing only too well of the young girls who were ready to bribe unmercifully for a sight of his charge: malignant, with a show of pride, he turned the best possible light upon the prisoner.

The floor was dirt. Through the high window I saw a cold star, then a few flecks of snow. The cell was short, the eastern wall a ledge, stony and down-sloping, upon which the prisoner was to sleep. A large wooden spoon and fork lay there and black thistles shedding pollen dust. At first the prisoner tried to protect his eyes from the lantern glare, then slowly he got to his feet.

"Mio prigioniero," I addressed him simply and said no more. He looked at me; and the prefect also, with Pucento, looked at me. In the recessed eyes was a worn pleasure and, in this fortress cell, expectancy. Here the prisoner was detained in Sasso Fetore's highest stronghold, a man with two hands, feet, and all the past we can remember, our captured image, a foreigner. He blinked less in the rusty light.

"Little Ginevra," muttered Pucento, kneeling, peering first at the prisoner, then at myself.

"Shall he take off his coat, Hangman, that you see him better?"

Every now and then the prefect shook the lantern as if to shake it into other focus. Nothing was the prisoner's but what was about him, little remaining yet all, the hair on his head, the gray of the skin around his mouth, the coat.

"No, Prefect. But unlock the cell."

"Donna. He is used to no one except me."

Pucento's round head, the round head of the prefect, the lantern's head were at my back in the doorway, and I faced him with nothing between us but the air which we did not turn to intelligible sound. He wore a gray trenchcoat that buried his legs; he was hatless, and on his face was an expression of wistfulness. His collar was damp as if he had been breathing quickly all his life, on the collar silver insignia of a skull and crossed bones. The features, the teeth pressing against the lips, the eyes which had failed in his calculations now lay pale and aged with pupils in semi-focus. And yet all about him was the smell of earth, as if earth had been packed into his helmet, ferns packed in his sleeve, and the buttons, catches, and chevrons had rusted away.

I took his two arms and lifted. They remained outstretched, and I took hold of his lapels and pulled. I put my hands on his ribs and felt with care high as his armpits, slowly, and he remained standing despite evidences of the prefect's hooks. There was a lump on one rib as if it had mended of its own strength. He endured this inspection: and the while his attention, his form of halted intelligence, was upon me as if to find some information for his own welfare in a gesture of mine.

I was that close to him. And did not intend to be so near him again until the Pentecost was past. Still there was not a word from him, only the accumulation of strangeness, the signs that he would never be at home in the cell before we removed him. Never again quite locate himself, he who had lost his battalion of all things familiar and banal, his comrades. I suddenly found that it was with curiosity I searched him.

Under the coat, hanging from one shoulder, I discovered the black grained map case, much out of shape since he had slept on it, and across the front of the case were loops containing two thick writing instruments and a steel calipers. This I took and opened. A thin sheaf of maps was tied inside, a packet on thin paper and wrinkled with constant exposure to water. They were of our province, the details, landmarks of Sasso Fetore recorded precisely, the perimeter clearly indicated, the place of the charred foot soldiers, the forest, and there was the fortress. I looked closely. Then handed the case to the prefect. There was nothing else. Only that he had known his way to us by these directions, the work of an old and shrewd cartographer.

The cell was narrow, the ceiling low, and the earthen floor was covered with the many sharp prints of boots wide and thin, pointed or blunt. The red light honed the bars, and the two bars also in the window beyond reach. I wondered what ruse he might try to get his head up to the window, what efforts he had already made. He was perhaps sensible enough to catch a glimpse of the night and to remember his homeland. Beneath the natural height of the cranial cavity were the skull and bones and the enveloping wrap of the coat with its accumulation from long roads, and the great collar which had protected him from the winds when he followed behind the laborious gun carriage of his century. He was the embodiment of caution, the human form endeavoring for obedience and sustenance. I felt the beating of his heart and in that instant he too seemed aware of it and ashamed of it.

Had the women in the streets seen him at all? And Signor Barabo, paused in the caffè, sleepless, he too had overlooked the submission, the fated attitude of the prisoner who might yet be beaten to violence.

"Altezza." The prefect interrupted me and handed me a document that had been carried with the maps. It was written in the language of Sasso Fetore, faded and washed:

If I have fallen into your hands, treat me with humanity. While in your captivity, I should receive half pay in your own currency. No outmoded punishment should be practiced upon me in the name of the Spirit. The agreement is that I shall not be maltreated. You who take me abide by the charter. I am to have no fear for your charity. I will not give up hope. Regularly give me water and a ration. Honor, publish abroad a notification of my surrender that I may keep my place in the world, even in the separated ranks.

"Primo Boia," shrieked Pucento from his knees, making some sense of the paper, "tell him the truth! Tell him he will receive no visitors but these!" Pucento, my ward, reached up for the document, and I gave it him, with its notions of temperance and gothic print.

I stepped back suddenly from the prisoner and saw him sacrificing his last days to conversations with insects, growing a beard, and fasting without recognizing hunger. He would study his fingers closely, the expression of consequence would pass from his face until he was carried to the piazza of the tall lady. I saw a curious crudeness around his mouth. I looked at him, at the shape of his jaw, his height, standing with all the turmoil of his senses guarded, his knowledge serrated, and the skull and crossed bones were his last insignia.

I stepped away from him. That one human might inspect another, I peered at him and was aware of the declarations and betrothals within him. But, as he raised and bent his arms, I saw only the white tips of his elbows protruding from the sleeves in the coat. Signor Barabo's son-in-law! He was ragged! I would not remember him for long. Nor certainly would Antonina. I would not see him in such calm again. And then I shouted at him: "What is your name, *immediatemente!* Your name . . ."

He did not answer, and we left him in the darkness.

Pucento and I wound our way bottomward, both of us silhouetted at the rim of the cliff and against Sasso Fetore, its obstructing roofs and chimneys through which no one dared to call the watch of the night. The last lock closed behind us.

It was late, the doors appearing in the darkness were not plumb. The nosegays of the welcoming crowd had been swept from the streets and they were deserted. The immense king's evil of history lay over the territory, and it waked me, as in the dawn, to breathe deeply, and I raised my hands at each doorway putting the seal upon them. Pucento careened ahead of me. Had it not been for the curfew, we would have been approached by those out of the dark begging their fortune. But the noose of night was drawn.

At my doorway, however—once we had passed under the arch and by the stilled Tuscan fountain—there was a disturbance, a fluttering and tangling with the bell chain, so loud that the old beast stamped in his stable. The prowler, come perhaps to intercede for the prisoner, was caught by the owl and, with fury and pointed ears, he sat upon her head, slowly raising and lowering his wings as a monk his cowl. He dug into her scalp, circumcising the brain. Her tresses were gathered against his dirty tail and he tugged as if he would carry her head up into the air. The owl labored and beat upon the woman, rasping through his gray hood. She tried to run, but he was fast to her and flapping, and the beat of his flooded wings slowed her.

I relieved her, taking the owl to my arm and comforting him. Pucento whimpered. "Little Ginevra," I said, "for your sake, you had best return now to your father." She fled and held her hair and the wounds in her wild youthful hands.

After a long while the owl's wings began to settle again to place, stiffly, with reluctance, the stimulation and traction, once summoned, loath to leave his wings and allow them to lie furled in sleep. They continued to preen and rise irritably as with the urge toward flight, ruffled, mobile, his mode of propulsion uncontrollable. But I wet and smoothed the feathers under the triangle of his beak with my tongue and he regained himself, once more folded into his nocturnal shape, and only the eyes did not relent. I gave him a large rat and slept near him the night.

And I dreamed of the universe of the tribunal. It was a

closed sleep and a closed dream in which the tenacity of elements parted layer after layer to spaciousness. I dreamed of a brilliant morning and—I was remote, standing away—I saw the three turrets of the fortress rising each from its peak, and the mountain of Sasso Fetore off there was a pale green.

From each tower flew a small white standard, constant and square in the wind. It was a dream of the three white flags which were suitably the ensigns of Sasso Fetore, starkly bleached and deliberately unadorned with the hangman's owl. Their white was mounted briskly above the green. The country was no larger than the flags and as perfect. The road was a bright red line winding to the three precipices and the capital of rigid existence. And the flags were moving, fluttering, the motion of life anchored safely to one place.

A soundless wind. Then some silent battery commenced a cannonade from a distant point in the light of morning—not a figure appeared on the battlements—and a silent invisible grapeshot tore at the flags. The white standards were pierced and began a silent disintegration until they were no more than a few shreds beating solemnly against their masts in the blue sky.

And out of the blue sky came Monco's voice, wily and cold across the plains and fortresses, concealed in the rays of the sun: "The fish are running well. The fish are running well, Master," with mockery in his voice.

THE PRISONER ESCAPES

"When shall we meet again, Hangman. When shall we meet again?" whispered Antonina, the belle, as if she would have the moment exactly.

Her body and her character were her contracts and she called me by name, Il Gufo. She had firm passions and firm words, and I felt the responsibility of her declaration made so clearly, and the question asked simply in the word of parting. *Appassionato!* She was white as the Donna these days and hourly grew taller as her love increased, and her passion was of moral seriousness to a woman who was herself a covenant. Even a woman such as she, descended from fourteen dames of wealth and modesty, might make her words sweet and sentient in her stand, the shape of her throat, the tone of her voice which was always of the clarity and deliberateness of the fire that could but burn so bright. Yet it consumed all there was. Absolute, dark-haired Antonina, her arms were thin, neither fairs nor family had trifled with her complexion. Under a green tree in a black field she read her sister the *Laws of the Young Women Not Yet Released to Marriage*—and in none of the laws was there heartbreak—while, her jaws moving, she cast inside herself the terrible beam of introspection. Hangman!

Quando potrò riverderla?

Little Ginevra was younger; Lucia and Teresa also were younger than she. They had not known the waiting. Their bridal purses were not as hard as hers, their heads not so high. Nor could they look upon the hangman and his equipment without shyness, and they aroused the owl. Antonina would have grown old loving the bleak rose campions and myself. She wrote no letters to the dead, but sometimes after I passed, she walked among them. One would have looked for her white gloves to reach around her darkest years, one would have thought to meet her at the green arbor. A stately, infatuated woman, she carried a love poison and a shawl for her neck and shoulders to the summer fair. At the fair she was met by Sasso Fetore's sisters, walking in pairs as if they had already ventured too far from home, coming from opposite directions bareheaded, unescorted.

Down the cliff they came, down the road like black angels, and I heard their whispering, the complaints of goose flesh on their arms. Hardly a woman or girl was left behind in the thin black streets of Sasso Fetore, vacated as they would be at the end of existence. But the field below was filled with the noise of feet that would kill the crabgrass, and I saw how few childbearers remained for all the pleasure they seemed to be taking and despite their demands for husbands.

From a clump of nearby bushes the old fathers of Sasso Fetore began to strum the viola da gamba.

It was now midday; the sun beat upon the field. Garlic was on their women's breaths, their appetites were sated with the macaroni that might make them milk. Little Gineva was there too and wore the matron's wimple that covered her torn skull. They stumbled, the swaying shanks of hair, the flaming red scarves binding torso or hips and the cantilevering, the maneuvering of the skirts. Newly shod and gowned, the purple and green of earth and sky became warm in their presence; all that was female, unnatural in congregation, came into the open air walking as geese who know the penalty awaiting the thief who catches them.

Antonina had dark eyes. There was no girl's foolishness in her bosom. She was accompanied by her mother whose great square cheeks were white as salt. One dark and one white they passed, a tall one and a short, posting their two figures against the hagbush overlooking the field.

The fair was pitched directly below the fortress, in good view, and for the benefit of the prisoner up there. The black schism of the fortress fell thrice across our ladies' heads. In the silence of the hilltop, in the window, was the eye of him alone who could not join them. And I too, at that hour, felt the urge to climb again into the city. The owl and the prisoner remained and the women showed themselves freely, their voices drifted high and were suddenly clear.

History had forbade the fair, a guise for flirting and the dissatisfaction of a sex—the fair invoked only when the measures of fathers failed. I listened to the festival, the ribaldry of the viola da gamba, the concert of bushes. How could it be anything but an ill omen, the distraction and the gaiety of woman preceded the fall of man. Women could not be quiet when men met and stripped back the skin from their arms and presided over the bare bones of inheritance. What a time for the tithe of pleasure, as if the sun must assert itself before eclipse and radiate before bursting into the fens of winter.

For most of their days the women were threadbare, garbed for seclusion in gardens that were high-walled and bloomed only with a few sullen leaves. These faces were not classic, but in the charity of the fair one suddenly seemed good as the next, a dispensation granted to each with a large nose, eyes that were too large, and manners that were not proportioned. Like a useless cadre, the twelve Mongers stood in a line at the south limit of the fair, the brims of the twelve black hats touching and inseparably joined.

"If it pleases you, look at my daughter Lucia. Her mother has dressed her in brocade for the day." And there were more and more daughters; they promenaded on this slope where the lay brothers had fought with fists, and rope girdles entangled like fighting bucks, so long before the season when the field became spread with green.

"Mamma, I will carry your train. Un momento," spoke the very young, having forgotten the goats in the stable. The proclamation was wet upon the north wall. They had gold rings in their ears and other luxuries hung to their bodies. At just this fair in the past some were got with fornication and games, in the time when there were men to hang and those to spare, with clemency for neither.

"Attento, attento!" started the amusement and in the almond color of noon Pucento the lictor came upon the field. The women quieted and did not crowd beyond the small white markers but watched each beside her neighbor.

"To the hind legs!" Pucento shouted and proceeded slowly toward the center of the green. His walk was stiff, slow, itself determined by the beauty of the aged provincial combat between man and dog, the mastery of training over the temptation and distraction that plagued the low species. The women peered from the four sides, some scowling, some thrusting forward their cheeks withered as nuts, some smiling as if they alone were devoting themselves to a glassy pleasure at the sight of my ward high stepping and lean. The young girls watched covetously.

"They suspect something," murmured Signor Barabo and held Antonina by the waist, confidently. And where he stood also stood eleven Mongers more, their wind-troubled sombreros cut at different angles from amidst bodices, old women and their daughters.

The tempo of the viola da gamba increased, but there came only the soundless winging of the musicians' bows, the silent press of the female hundred, and the stilled orderly panorama of the fair. The olive faces, the roses, the chameleon breasts were ranked and slightly moving under the hemispherical silence of Sasso Fetore. The noon tilted overhead. Antonina seemed not to feel the binding of the arm about her waist and did not watch Pucento for long but looked directly away and to the east, steadily. "Adesso," came the movement, the will of her lips.

When the dog turned, they turned, and the oldest and most grudging, with lace upon the brown of their chests, paid attention and swung the great fans atop their skulls windward, frowning through malignant black eyes as if they would not be fooled.

The streaks of the gowns against the earth, the moving flecks of the man and dog, the liveliness of the noon and the windy pasture below him were the last that he in the fortress would see of mankind, womankind. Each minute the grade up to the fortress became more steep and the music only a toneless drift from the strings soon to die.

The dog was muzzled. To the tip of the muzzle into a ring fastened at the end of the snout was hooked the leash which Pucento and the animal kept taut between them, a thin rein of brilliant red. Pucento held his arm, the fist gripping the cord, straight as long as he and the dog maneuvered together and the one obeyed the other. Pressure about the head controlled the animal; two leather cups on the muzzle hid his eyes.

On their leaning instruments the musicians played the seldom heard 'March of the White Dog.' This whole breed had once been deprived and whipped, tied ascetically by the lay brothers on the slopes. The bitches were destroyed. And the rest, heavy of organ and never altered with the knife, day after day were beaten during the brothers' prayers, commanded to be pure unmercifully. The dogs tasted of blood given in mean measure but were not permitted the lather, the howl, the reckless male-letting of their species. Beaten across the quarters, they were taught by the monks the blind, perfectly executed gavotte.

The sole remaining dog moved and balanced as the first packs, flawless, the long wail of refusal still in his throat and still denied him by the muzzle. The dog followed Pucento on the end of the tight rein, a heavy animal, the white coat become tarnished and cream with age. The women could not see how Pucento sweat, himself straining to duplicate the measure, the ruthless footstep of the past. Welts were knotted across the dog's hide, causing the hind muscles to be tough as if the leg's tendons themselves had been drawn upward and bound across the spine. The joints were round, distended, polished to silver, thick though the legs were slim, worn and delicate with the hours of balancing. But the meat, the shoulders and loins—tempered by the monks—were broad and considerable so that the animal might endure the requirements and travel long distances without touching the front paws to earth.

The old dog did not once attempt to snap through the muzzle. It changed the rhythm of its gait perfectly and moved sideways with ease, crossing one set of claws over the other. All the desire, the reflex to kill, was still there under the white coat, inside the white skull and embedded the length of the spinal column, but Pucento had no need of the thick cane, no need to thrash the animal in the formality, the difficulty of the devilish dog's fandango.

First the two completed a square, then a circle, then the dog twisted and arched its back. Drawing down the quarters, it puffed and deflated its chest in one place, the leash pulling always the primitive long jaws and the restrained skull horizontally forward uncomfortably from the neck. But in this position the animal's silhouette was best.

"Attento, attento," they murmured again after silence and Pucento stiffened his arms, man and dog frozen on the green. Pucento spoke. The dog shuddered and swung backward in brutal symmetry, lifted, stood on two legs, then leapt, once, thrice, and each time a single leg only touched the earth, quivering, burdened, unnatural. On the one leg, the dog propelled itself upward again without falling, and the bones pressed through the fat. Even Antonina's mother did not regret the sight.

The dog stopped. As bidden, its front paws came to rest lightly upon Pucento's back on each side of the neck. Thus they remained rigidly, heads damp, white, lifted into the sun, while the musicians dragged their viols onto the field and the girls raised and silently shook their clutches of rose campions.

"Antonina," said Signor Barabo, "take your sister's hand."

But Antonina was gone and his arm was hooked only about his old wife's hip.

We climbed rapidly, Antonina and I, clinging to the steepest ascent. We pressed ourselves into the declivities of the cliff. The undergrowth, as it scraped the hand, was warm; now and then we moved upward through the devil's mace and were stung by the lonely nettles. Antonina took the path first and did not pause. She seemed to climb with her narrow shoulders, and there was spare straight movement under the twelve skirts. The music from the fair still reached us persistently up the bed of a mountain brook, la, la, la, la, so that we hurried. The women below trod across the green, a few disappearing off the edge of the slope.

Who has not wanted to climb on a warm day, up again toward the bare hills? We passed without thinking of the trickle of dispassionate water from the fortress. The air was clear and almost free of Sasso Fetore's garnishing odor of rust and yellowed tomatoes. I drew closer to my companion. Antonina made herself known, and we climbed again.

The whiteness of the underskirts lay against the rock and

coils of mountain grass. I heard my own boot slip and start, and I was behind her lest she fall. Antonina's pale hands touched the calcium encrustations of the rooks, and the wind took her face and clothes as if she had mounted those leaning steps to which the faithful will not return. Perhaps in her heart was good conscience for all her years.

Only that morning, so soon, she had distributed the contents of her bridal casks along the embankment to sun. After sunning them, she had made several bundles, tied them with cord, and carried them up the wooden ladder to the dark space beneath the roof, knowing they would not last there and that she would not need to take them down again. Her father would not speak for her.

And yet the thorn pulled at the leg, a trailing of her shawl was snagged in that steep place below the fortress. She swayed and proceeded to climb as if there would be more trysting. We were hidden by the glare reflected from the cliff, high where not even sheep grazed. In her hurry, her determination, she moved as if to absorb her indiscretion into the blood of her good family. Now she laughed.

At last we reached the ledge and stood side by side, then face to face so that I could not mistake her, Antonina. Not from weariness she leaned against the lowest walling of the fortress morticed agedly into the cliff. Already her breast was rising and the noon fled. We had no need to whisper, not even the birds were within rock's throw. But the wind was in our faces and we were temporary, though Antonina did not look as if her heart were sinking. The world this high creaked around us and, standing with no sure footing between the day before and the day after, she touched her bosom done with lightheartedness, spoke to me in the wind's way:

"Honorable Hangman. Carino. Il Gufo. It is you I love. I know what women do, and I have no fear of it. I have heard my father. I will be no belated bride. 'Not him,' they say. But it is you I love. I have seen you ride your jangling ass toward the rope readied and hung down from the sky like thunder. I remember the superstitions; I am old enough to remember them and you. 'Not him,' they say. But it is you."

And Antonina held to the leathern flanges on my hip, there on the cliff, among reed and empty eggshell. The fair was done; the waiting of rags done. I put my hand on her bosom and my hand met the two small silver hearts of a fine lady.

Antonina rolled stiff on the brown hilltop and the skirts loosened, lifted by the wind. She pushed her fingers into the bent grass and dragged her hair on the silt and stones. Her slender belly thrashed like all cloistered civilization among weed, root, in the wild of the crow's nest. I reached into the sheltered thighs touching this bone and that and felt for what all women carried. High and close to her person, secreted, I found Antonina's purse which she had hid there longer than seven years, that which they fastened to the girls when young. What was there more?

There was the prisoner. Having found his way from the room of four hooks, through the base passages, he climbed until he could go no further, bent and blinded by the light, clamoring into the air and to the stone above us scaled by no ladder or foothold. Now he freely cried *Guai!* Lifting his roving eyes away to the roofs, the spaces through the city, he seemed to fill his eyes with the distance to the borders. His was the rage of prisoners who climb quick as they can to the rooftops, who are caught in the tall trees—there was hardly that to steady him or give balance—reaching at last this windy free space. Perhaps the fair, the sudden quiet, the loneliness, made him understand that he could not escape the way he came. He was transmuted and prepared for the dizziness of the high ledge; the sun, the air currents, caught his face. I looked up at him and raised my hand to hold him.

The prisoner was covered with great feathers, pin feathers and flat feathers, pieces of wire and tin swelled his chest. The wings hung far down as arms and even below the hands, swaying, and were fastened across his shoulders. He crouched heavily, but his waxen feathers, his flying skein billowed angrily in the wind. His head stuck over with red wax turned loftily. Then he tested the wings, looking at the sun unbelieving, taking a cautious step closer the edge. The wings hung down and buried the arms inside; almost to the length of his feet, the tips waved like the lengthy, extra feelers of the dragonfly. The ends of the wings were wet, they motioned under the power of the primary feathers, the crudely fashioned wing with its sharp trailing edge. And when he filled the wings, they moved, lifted once, again, curving down and menacing. His half hidden chin jutted and thrust with the effort. He spread his legs and drew tight the red flying surface between them, so that he was a mass of machine and bird for the wind's picking. He appeared heavy as stone.

The wings caught, and he burdened the wings, the wax, and the red cowl from the rusty forehead. The skull and crossed bones were buried under the brown breast. He sapped the wings and his shod heels lifted, the knees flexing and ready to hurl him off. The eyes grew small in that headgear, birdlike, as if free they could distinguish only black and white and the long distance, in any direction, that there was to fly. None before him had thought of it, none fabricated such a means of escape. And his head raised, the urge to leave the earth and gallows, the very hilltop of Sasso Fetore, lightened the drag of the feet.

He tore himself away. He poised himself on the great stone and tried himself, peering aloft and away for some landmark by which he could travel and survive. The face was criss-crossed with red lines, he had discovered even the crop of the bird. The tail—for he would guide himself—spread out. Still clinging to the stone he wheeled once, then back again and pulled the feathers, not hesitating, merely tightening the tufts in the wings and drinking at the air. Behind him I saw the top of the low tower and the blue atmosphere. The wind blew up stronger and clear.

The wings beat slowly down. Then as if to break the bones in his arms, they were horizontal, sweeping a little windward. He brought his knees into the pit of his stomach and climbed toward open sky. The prisoner hovered, turned awkwardly, swooped close over our heads—he kicked the air as he dove!—and sailed in a long arc up again, around, about to disappear across the witch's huts and chimneys of Sasso Fetore now darkening with the night already close. I saw him lastly fly defiantly through the smoke of Monco's deserted fire and into the red sun that sank and drew him down over the edge of the earth.

HE HANGS

Finally the night enforced quiet. The tunneling cicadas, the cicadas taking the moonlight on patches of snow, were still. The shadows maneuvered, the eternal flanking movements and frozen sorties of the night were taking place. The brook ran cold, and the mountain slopes, so far and of silver, marched upon the rain, absorbed the moonlight like black cloth the sun; houses were deserted at dark when the four-legged animals, not many, hung their heads. The fortress imprisoned only the empty hooks. The ice span told no time. There was the webbed fern, the rafter, the proclamation's promise, and the iron doors ajar—the bishops and gaolers done convening and taken to bed until morn. The galax opened its sharp leaves to prick the prowler, and there was an odor of night's roots.

The owl was awake, he swelled his chest, breathed restlessly, and made himself known in the dark as if it were not deep enough. He was dissatisfied with the still lingering light and kept to his corner. Now and then, slow and purposeful, he eased himself from the pitch and moved in front of the window, his claws biting the ledge, his outline contesting heaven and plain, tree and pillar. He looked and the bare window was driven behind him, or blinked and was awake, frowning at the fortress and murmuring skies beyond the cliff. He looked so violently he heard nothing. Insects darted in and out of his feathers and they were nothing to him, so long did he stare to see that all was in the night as before, and before that. He looked steadfastly at the universe, then turned his back and proceeded to chew, pick, beat his cold heart, rustle so imperceptibly active with continuous life. Revengeful was he toward that which made him turn his white face and look into the dark.

He was old, scabrous at the window, he regarded the night from his stone and branch and all the night was preoccupied with some stretching of tissue or memory deep within the feathers, while rectifying the vision of the world in his owlish eyes, watching it as he might something that dared not move. He silhouetted himself and from that glance the night could not shrink into hiding in the atmosphere. He gripped Antonina's purse in his claw and now and then shook it, already it was ripped and musty as if it had been his forever. At times it fell to the window ledge and he kept near it.

Back and forth he went, continuously at work, conniving and busy within his feathers or lice eggs, watching the dark, flitching, flickering. He cut his bill on the stone, preened disinterestedly. Once, with slow effort, with a great plying of nerves and muscle he slowly shut his eyes, down, down, and obliterated the shadows, the space, and it ceased. Then, look, the dilation and they were round again, greater, the horned moons. He moved across the rim of his nest, Sasso Fetore, at the window. He outlined himself again and, face into the night, his head began turning so little, but sharply, right, left, and this was his alertness, something of all he saw aroused this speculation, something fallen upon that eye stirred him.

The owl carried Antonina's purse to the window and shook it. And the leaves in the scrub tree shook also. One claw was missing, another cut off blunt and short, at every feathered layer he was scarred and covered with old wounds that penetrated him like the grain of wood, his fiber, the old markings of the forest. But he had never cried with the pain, the scarred face, the face enraged and bloody, always anesthetized with the cold enormity of the eyes, the sudden circumference of the eyelids protecting him, making him insensitive.

Light began again to rise steaming from the earth, and from the owl's eye, slowly, lighted the knowledge of the day to come. He stiffened, watching the cicadas stop, and all about his motionless body the hoary piercing feathers rustled in the breeze as if they were no part of him. He glanced at the frosted shadows, the warped tree laden with a winter provision of dead mice, at the cold pump. It was morning, but only he could see through the blackness yet heavy to the light that was kindling. And he hooted, warning that someone was approaching up the winding road.

Signor Barabo came from the mist and halted near the window. He was wet, having come at this odd moment to take the owl's attention.

"Il Gufo. Boia?" he whispered, stooping and white. He looked up and bowed his head. He on the ledge waited. Signor Barabo stooped under his burden.

"I had not expected, Master . . ." turning his mouth on its side, whispering, squinting out of the frost. "For her I could not have asked you. Boia . . . Barabo's good fortune, I would not have missed it. Principale, I would reach up to you. . . ."

On his back he carried a black trunk, rounded and banded at the ends. Wet black hair fell from under his broad hat, the face of whiteness and dark lines hung there, smiling, the neck bulged with labor. He had walked all the night.

"You must pardon me, Altezza," grunting, wryly twisting under the load, "two men could not be found. So I have come alone. Half the treasure is here, Boia." Once he raised his voice, "Dowry," he exclaimed. He stood at the edge of the cliff and his fingers clamped the brass handles. The black lines of the face, sweat-strewn, slyly peered upward. He had discarded the cummerbund and his ankles were thick with mud. "It is a pleasure, Boia, for me to bring this heavy load myself." Since sundown he had carried it.

And suddenly he dropped one shoulder, swayed as if he could bear it no longer, and the cask was loose, fell out of sight and split wide on the first rocks, smashing, and its collection of coins showered down the slope until all was quiet again.

"Ah, sfortunato!" His eye cut narrowly and he wiped his nose. Signor Barabo turned on a heel and, his voice coming as through the rain-drenched forest, "Hangman. Follow me." A gust of wind carried his hat also down the cliff, and with that, again he fixed upon the owl his night-worn face.

Pucento saddled the donkey. We set off in single file down the morning rim of the quarry of Sasso Fetore. Pucento walked at the donkey's head, limping, pulled the damp halter. All about was the stillness that follows fugitive action, and we proceeded through that time of dawn when the werewolf gives up his feasting and the assassin lifts his hands from the jugular vein of history. We met the first citizen in a doorway awaiting our approach.

"Triumph to your day, Hangman," and the citizen fell in line behind us until we met the next, shivering, and more after that, sentinels suddenly coming alive from the low entrance to the caffè and along the down-winding route, those who with their hands and feet, their persons, making no noise, pointed the path that we should take.

"Is it the same?"

"It is, Signor Barabo."

"They have not wakened perhaps in my absence?"

"They have not wakened, Signore."

So Signor Barabo questioned each. These men, as they swung one after the other from niches and cold casements, out of the walls, followed us scenting and long-legged and sleepless, wet with uncomprehending vigilance. Over their brown shirts they wore short black jackets strapped high to their white throats.

"Power this day, Il Gufo," they murmured, surprised that I had come, and were gladly relieved.

The donkey braced his feet on the stones. It was not yet morning, the bells marked time coldly. No one peered from the windows, none proffered fowl or wine as we passed in single file. A gargoyle hid its face in corroded hands. I marched down through the people's hutch, thatch, sleep.

"Who lives here?"

"An old woman, Boia. She is asleep."

Even I had not seen this yard before. Under the hollow tile was an iron wall disappearing without vine or air hole into the earth. Leathern feet crowded and stirred upon her property while she slept, and the gate to her yard was open, pulled from its wooden hams. One had mounted the stones-avalanched between wall and privy; silent, motionless over the others' heads, he watched the approach from the hill below. Another with large stride paced off the width of the yard and tapped his head grimly to remember it. There was tarpaper over the old woman's window, a trough filled only with a bed of fermenting straw, and the back step where she took the sun, all become cold and removed, visited by the dawn Mongers and the inquisitive strangers, because of the event perpetrated on this spot.

The deputation filled her yard like black chessmen thrust into the dampness and sands of ruin. The bony ponchos and black hats hovered to and fro, staring at the corpses in the idle chill.

The prefect was on his knees. His eyes protruded, the kepi was at his side. Signor Barabo sat down near the prefect on the trough of stinking straw. Traces of light began to flow onto the horizon with the roar of wind.

"Can you tell when it occurred?"

"Perhaps, Signore. Perhaps they have been here a few hours. Or more."

"And they are dead then?"

"Dead. They are cold, Signore."

The prefect's four ganders lay at the base of the iron wall and their white bodies were frost covered and gripped in the weeds, to be seen at the prefect's knees with all their dismal inertness and roundness of breast sparkling and portentous. They appeared white through the feathers, through the flesh, and to the earth. Their eyes lay like black berries half in the sand; through the bills the nostrils were bored with blue augers as worn holes in wood.

"Four hours at least, Signore."

The long-lying feathers were immaculate, though the wing of one bird was crumpled—fanning, uncomfortably jointed—against the black wall and hung there hazardous and spread as if it would defy the upsweep of air, no longer temperate, in Sasso Fetore. Otherwise, in the small space, they geometrically marked the four points of the compass, unruffled, exact creatures as they were left after death. The prefect did not touch them, he merely leaned further forward, put his hands upon the dirt near their circle and bowed so that the smoke from his cigarette burned his lip. He peered at the ganders who would no longer invade the impossible cliff top and campanile of Sasso Fetore, bedded down now emphatically in the place they would not have chosen to protect themselves from storm.

Not turning, with the smoke still branching and forking back from his lowered face, the prefect whispered: "Do you see the murderer?"

The one on the rocks hesitated. He opened his eyes wider, examined quickly the architectural slope and the ferns, and shook his head.

Signor Barabo's mouth fell as if to exclaim, but he remained silent. And one after the other the black figures rotated, stood a moment at the prefect's shoulder to look.

The ganders would no longer march through the steep places of Sasso Fetore, circling upon ridges, gables, spiked walls. No longer would they search the flagons—search the tall lady's piazza, the fortress, the field of dead foot soldiers—for the cliff and kernel of Flemish seeds, for a crack in the ice, for the remnants of dark days.

"They are dead, Signore, I feel it here. These are the lumps in the neck, there is a hardness under the bill, Signore."

And the prefect withdrew his hand, spitting away the smoke and tobacco.

The ganders had been felled carefully, symmetrical and clean. The long necks, straight, each perpendicular to the next, were crossed one over the other near the heads, the necks touching and torn, left in their severed lines and with their cold windpipes in this intimate, unnatural pattern. The ganders, whose eyes gleamed logically, whose march was rhetorical, surmounted the prisoner's sudden inspiration to kill them and survived his warped and cunning urge to lay their bodies in crudely artful fashion.

The donkey rang his bells dumbly and none heard. The straw fermented. The guard on the rockpile lost his eyes into the mist and gave no alarm. There was no sign of the one who fled. Pucento, with tentative finger, felt over and over again the slashes imprinted on the earth by wingtips and fierce talons. Another stood waiting, and, from the wooden saddle, my feet slanting as iron and one fist upon the tin tattooed horn, I motioned them away, signaling that they leave their posts and wet odor of the yard. I pulled on the rein sharply.

The dawn became the color of the pear's belly. And not long from that time the prisoner's discarded, tall, half-broken wings were found abandoned against the whitewashed sidings of a stall. The straps that had bound them to him hung stiffly. In the mud his footprint was recognized where he struggled free. Before the Sabbath labors began, the citizens of Sasso Fetore saw for themselves the old woman's yard and filed past to stare at the wings by the barn where they were rumored to be. When darkness came again, the wings were illuminated with the flare of two torches and were visited even then. The children looked for signs of his toes in the wet sand, they whispered upon hearing the bell that was struck hourly as long as he was free.

And he was not free for long but was returned to the cell from which he had escaped, to the four hooks and put upon them. No one gathered at the fortress, satisfied with the announcement that he was there. His wings were splashed with kerosene and burned where they stood. In a matter of hours his cries commenced again, and peaceful anticipation possessed the owl. The citizens listened quietly as they stared at the pool in the grass, now empty. And the sun set. The crowd dispersed from the smoking wings that cowered, withered, fell to ash. Some days later—the prisoner was still suspended on the hooks—the skin was drawn away from his belly in one piece and stretched across a drum that was beaten through the streets while they stopped work and listened.

The Pentecost was past. The prefect, as was prescribed at this time, fasted and shaved his head. He was ready, imprisoning himself with the condemned until he should be brought to me. All of Sasso Fetore returned to read the proclamation with more care. The drumming upon the prisoner's flesh continued, sometimes the drummer was followed by women and the Mongers, sometimes he beat alone through the steady rain, and he took the old courses of the ganders, going nowhere in particular but marching and tapping the drum so that all would hear during the day. Sometimes the cries stopped and that too was heard. The old men waited in the caffè and they no longer spoke of Lucia or Teresa.

"Listen, Signor Barabo. It is past."

"Yes. And listen, thou, do you hear him hammering?"

"Yes. He is coaxing his noose and knot."

"Il Gufo."

The owl waited for the drummer to come, and each time bestirred himself so that he might watch. He sanctioned the herald to stop and to strike four extra beats for the owl and move on.

The millet grew ripe and hampers of blood apples were left in the hangman's piazza. A stone of bread was left also. Those in the caffè waited and made no move to kill the lizards that came from their caves in the fireplace; the bent grass sprang up again near the wall in the old woman's yard; the rooks flattened themselves in their nests on the cliff. The fox came into the streets but was not noticed, traveled up and down my road. Pucento brushed the donkey's black trappings, the monarchial owl stared about the proximity of the execution. The prefect buried the ganders, and Sasso Fetore had nothing else to think of, nothing else to prepare for.

One morning the drumming ceased. The fox hastened back to his hole. The bells ceased. Monco did not fish that day, and the fish remained deep in their pools. The cicadas were hushed. The cries were hushed in the fortress, the prefect giving the prisoner water from a tin cup. That morning Signor Barabo saw nothing of his wife and daughters. The fortress towered its three parts over the city and nothing stirred.

The yellowhammers appeared and flew into the sun, and the rose campions speckled the hillock, the path by the headstones. Sasso Fetore was raw with the sun that fell headlong upon the unoccupied streets and before the campanile, the waiting city. Bright, clear, reddening the cliff, the sun rose and spread down across the unlabored fields. There was silence. None came from their shelters while the shot of the sun tore at the white flags raised on the bare masts and shone upon the ascending slopes with their briar, a few huts, inviting the spirits to come again to that empty plain. All day the sun warmed the copper in the eyes of doors and dried the tall lady in the piazza.

"Listen," murmured Signor Barabo, "it has stopped." And they listened, not venturing from the dark. It was a day during which the citizens sat until they must change a knee or arm and whispered while the sun swelled the streets with the light. No meals were prepared, but now and then they drank small quantities of spumante or rosolio. They had not forgotten the long, white, and crossing necks, nor perhaps their daughters. They had not dared to right the Donna. The Mongers themselves could not escape the superstitions, even when the sun whetted the gardens and filtered to gleam upon the wine presses.

The crabgrass grew again. The old men did not attempt to seek their neighbors; those who were already congregated when this day came remained so, waxen within the cellars or the caffè. No one gave thought to attiring the young women in hammered silver and lace. The council of crisis and the occasion of tragedy reached them during daylight hours and when the sun was high, Sasso Fetore was lighted the whole day and filled with the ferment of the sun rays. The day passed.

Only at dusk did a soft murmuring, the dialect and talk, start from the houses. Then some tried to remember the prisoner's face and they haggled over who had touched him and who not. The old woman who had proffered him the roasted fowl declaimed that she had retrieved it whole again, that he had not a chance to bite into it. The fathers would not speak a word to their daughters.

When it was dark Pucento carried in the kegs of grain, the bowl of blood apples, the pail of dissected chicken, the spice and paste and fish packed in leaves, from the piazza. He built a fire using the tinder and flint intended for the Donna. He cooked until midnight and heaped the brown pieces of meat and the strips of white fish on the stones and the fat drained into the fire, crackling and hissing. He brushed the earth from the blood apples and steamed them until they were soft and strong of odor. He cooked black twigs that tasted as chicory, and the water boiled in the pan, the bones fell from the fish, the skin and meat became tender and dropped loose. Now and then he returned to the court and discovered another homage food deposited there, perhaps a vegetable the color of rattan. And this he cooked. He found a bucket of eggs no larger than walnuts. He boiled them. And he replaced the eyes in the fish heads. He stirred, snipped, built up the fire, and we drank cognac. One small bird he roasted and this he stuffed—showing his teeth and pushing his thumbs—with a handful of green grass pale and fresh. He burned his arms and did not notice, he worked his knife on the block and he bled; the blood apples broke their skins and he took them from the rumbling water with his burned fingers.

How different from the Mongers' fragile fish or the square dark loaf and water! The pewter was filled, the flat leathern platters smooth as wood were filled, and the chicken stones, the sticks wet with fat, the hearts like cherries and the joints, the twists of bread, were devoured, heaped up and then eaten. The owl sat long with a large and savory chunk of white meat in his bill. The smoke backed into the grate and was filled with the moisture of feasting. Pucento scooped at the black fish roe; I ate the sickle-shaped sections of a fruit and the fire lighted our red hands. Pucento sat with his elbows between his knees, his cheeks wet and his eyes heavy. Until the meal was done and the intestines could not move.

The donkey's freshly polished harness hung on the wall; the casket containing the rope was ready

"I shall see little Ginevra, Gufo."

Yes, and the rest of them. And the meal, torn from the anatomy of conscience, sat upon us, from the quantity found and cooked so seldom there came the effulgent memory of execution, step by step, dismal, endless, powerful as a beam that transudes our indulgence on the earth, in Sasso Fetore. There was suddenly the morrow, the way the brain sees death suddenly, and there was the penalty that could not be stayed. Lastly we chewed mint leaves.

In sleep there were the fantastic shapes of the food, the sensation of looking forward, searching ahead into the rubrics burning upon the slopes of the rose campions. Gathered about my bed were the hangman's articles. The gloved creatures of dreams paraded all night and a hundred times I settled the noose before dawn, and straightened it and turned it.

And the eye opened to find it cold, partially light, silent; the windows were open, all pushed a crack or flung wide, and the door was off its latch; the cold had come early and lay about the rooms. I rose, strapped a wide belt around my flesh. The stones were cold. I saw the ledge, empty and small, wet, from which the prisoner jumped. Pucento slept in his clothes and the lips were pulled back from the teeth. I shook him and he waked, clutching at my chest, and said: "Is it over, Il Gufo, is it finished? Have you hung him without me?" Then leapt from the bed and brought my cap and cape.

The crowds were in the piazza before us, ranked on four sides of the tall lady and pressing close to the gallows of white hickory. The empty arms, the cold hands, were folded and skull touched skull. I climbed to the top, the platform, and Pucento crept to the bottom, into the small space where the prisoner would fall. My footsteps crashed above Pucento's head and down there, out of sight, he crouched back against the wood and hid his face, trembling, sick as upon sea water, cowering in the darkness lest he be struck by the descending feet.

I stood. I might only have been searching the countryside from the low tower or kicking a splinter of glass down to the rooks. I looked over their heads through the slits. The soft black horn atop my hood curved forward and shook slightly in the wind. And the red cape, the collar of red, was short and left my arms free. The gallows faced the fortress. Not an old man, woman, and child but what was here and stooping in his brown shirt, gray cloth, and thinking of the book from which their hangman knew the terms and directions, the means and methods to destroy a man.

The wind blew and there was the odor of the hemlock at the border. The bare firmament of the cliff was hidden behind the heatless crumble of Sasso Fetore. The cold and dripping hole waited in the tall tree. My ward, of all of them, was sick under me. With my fists I tore the eyeslits so that I might see further and see them from such dilated eyes. I passed, shaking the horned hood at them and stopped.

The prefect brought the prisoner quickly from the fortress, under the portcullis, down the rampart, through the winding street, past the colonnade, between the crowds which opened, and into the piazza at last to the tall lady. The one and the other walked quickly and with single purpose, hair blowing, one official and one man walking as if they might not reach the gallows after all. The prisoner's greatcoat was open and it beat about his ankles; he walked now, came foot to stone.

And then there was no further for him to walk. The two of us stepped to the center of the trap, that board which shall be fastened so that it be firm at the proper time and fail at the proper time, leaving the foot nothing but emptiness. The muscles at the corners of the prisoner's mouth were hard like welts. There was not a minute more. He made a gesture as if to remove the greatcoat. But I bound his wrists, strapped the ankles, pulling the thongs tight. I put the hood upon him, down over the ears. I lifted the noose—higher, with both hands, lifting—and fixed it more easily than I dreamed. Then he stepped off the trap.

And that one noise of machine and man echoed and rebounded against the four sides of the piazza, against the campanile and the low tower, and disappeared down my winding road. Then in the silence the trap banged several times like a door. I stood at the edge of the pit where the rope before me descended quiet and taut, tugged steady as some line dropped through a hole into the center of the earth.

And in the crowd, back near the wall, I saw Antonina. Already her hair was gray and her complexion altered, the lips compressed, the temples shiny, and her habits and her character so true and poorly tempered that no man would come for her and the rest of her days would be spent with the manual for the virgins not yet released to marriage.

And then there was the air damp and cold and the owl exerting himself into flight, beating through the top branches now, shutting his eyes and crashing through the twigs to the dripping hole in the trunk, settling himself and sitting inside the bark for summers and winters, and he stayed thus, peering out of the warmth, the tenure of silent feathers in a cold tree.

Thus stands the cause between us, we are entered into covenant for this work, we have drawn our own articles and have professed to enterprise upon these actions and these ends, and we have besought favor, and we have bestowed blessing.